The

5G

User's Guide

The Next Generation Revealed

Jyrki T. J. Penttinen

Atlanta, GA, USA

Please follow the author's 5G blog at

http://www.5g-simplified.com

Contents, figures, and cover design: Jyrki T. J. Penttinen

Liability statement: The aim of this publication is to help readers understand principles of the subject based on best efforts in summarizing the information. Regardless of all the careful work, including research, source studies, and educated statements, there always remains a possibility for errors and misinterpretations. Before any conclusions or actions such as a use, development, deployment or operation related to the described system, the reader is advised to ensure the correctness of this information from relevant standard bodies, authorities and other specialist sources. Thus, the author, or any other party related directly or indirectly to this book, will not assume any responsibility for possible health issues, loss of profit, or any other damage based on the contents of this book, and disclaim all warranties. The contents and statements presented in this book are solely of the author and do not necessarily represent the views of the current or past employers or any other entities. Furthermore, any organization, website, or product referred to in this work as a citation and / or potential source of further information does not mean that the publisher and author endorse the information or services.

ISBN 979-8-6012-8596-4 (paperback)

Library of Congress Control Number: 2020901533

Imprint: Independently published

1st Edition

February 11, 2020

Atlanta, GA, USA

Table of Contents

The Author

Dr. Jyrki T. J. Penttinen, the author of this **5G User's Guide**, has worked in mobile communication industry for mobile network operators, equipment makers and security providers in Finland, Spain, Mexico and the United States since the first generation of mobile communications. His activities have related to the system and architectural design, investigation, standardization, training, and technical management. He has also authored a variety of technical articles and books.

Since 2018, he has worked in the GSMA North America's technology management team assisting operator members with the adoption, design, development, and deployment of GSMA specifications, with the special focus on 5G.

Please find author's 5G news and publications at:

▶*http://www.5g-simplified.com*

▶*www.linkedin.com/in/jypen*

▶*www.amazon.com/author/jype*

5G for All the Interested Ones

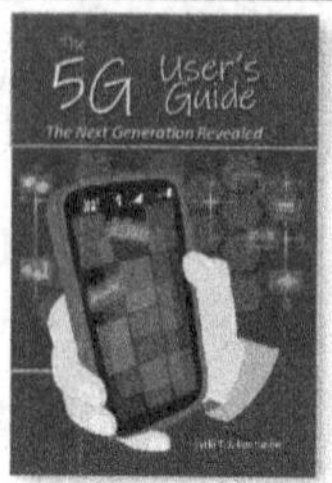

5G User's Guide: The Next Generation Revealed, Amazon 2020

This practical guideline presents the most important features of the new generation mobile communications. It gives a common-sense overview of the 5G system and instructs the reader to select the most adequate 5G device according to the need. This book does not require technical knowledge and is designed especially for the consumers wanting to know the basics of 5G and how to take full advantage of the new services.

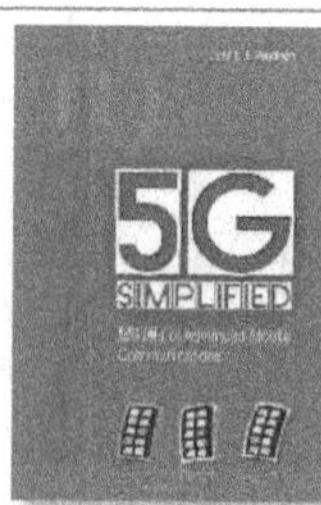

5G Simplified: ABCs of Advanced Mobile Communications, Amazon, 2019
5G simplificado (Spanish edition), Amazon, 2019

This book presents the technical foundation of the 5G system. It gives an overview of the new system's architecture, and goes through the functions of the networks and devices. This book is designed especially for the initial telecommunication courses and personnel of the telecom-related companies who are interested in a non-complicated technical introduction to 5G.

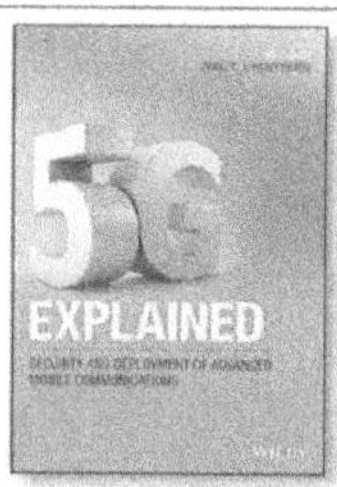

5G Explained: Security and Deployment of Advanced Mobile Communications, Wiley, 2019

This book details the technology of the 5G. It deep-dives into the 3GPP Release 15 specifications and "translates" them into more understandable technical descriptions. This book is designed for the mobile network operators and other stakeholders working in the 5G ecosystem to better understand the details of the architecture and functioning of the new networks, with a special focus on the security aspects.

Other Telecom Books of the Author

  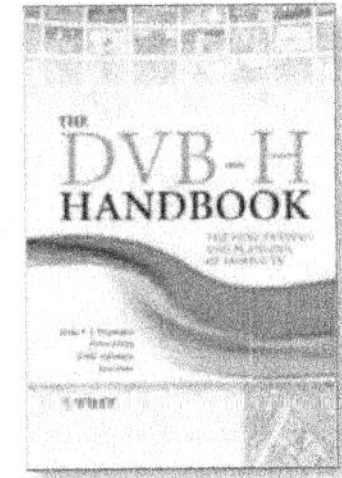

- ✓ *Wireless Communications Security:* Solutions for the Internet of Things, Wiley, 2016

- ✓ *The LTE-Advanced Deployment Handbook:* The Planning Guidelines for the Fourth Generation Networks, Wiley, 2016

- ✓ *The Telecommunications Handbook:* Engineering Guidelines for Fixed, Mobile and Satellite Systems, Wiley, 2015

- ✓ *The LTE / SAE Deployment Handbook*, Wiley 2011 (Mandarin edition in 2013)

- ✓ *The DVB-H Handbook:* The Functioning and Planning of Mobile TV (Penttinen, Jolma, Aaltonen, Väre), Wiley, 2009

Forewords

5G is about to take off. The new, fifth generation of telecommunication systems is superior compared to any of the previous ones since the commercialization of the very first generation back in 1980s.

While the technological base of 5G is still under construction, many stakeholders want to expedite their commercial offering even if the current 4G LTE networks may be often quite adequate for today's needs such as multimedia and video streaming.

While the industry is advancing fast with the new network deployments, we ordinary users may wonder what the greatest benefits of 5G actually are. Similar doubts have been expressed about every new generation, but like always, we consumers get used to the better performance along with the new data-hungry services.

Regardless of the availability of the first 5G networks, it is still challenging to capture the very key benefits of the new system from the buzzwords such as network slicing, virtual network functions, and low latency services.

This book provides answers to these and other essential questions about 5G, such as what is the best timing to obtain a new phone, how to set up the device, what are the main features of the new networks, and how to take full advantage of 5G.

The focus of this book is on the practical aspects that we consumers want and need to know about 5G. The 5G Revealed is designed to serve as a "survival guide" in order to use the 5G devices and apps in the most efficient way without any need for technical background.

While this book gives practical instructions, the interested ones can find more details about the new system from my *5G Simplified* book, which gives an easy introduction to the new system, architecture and functioning, and the *5G Explained* book, which is designed for mobile telecom professionals wanting to update the knowledge more thoroughly.

I hope you'll enjoy this guide and that it helps you to experiment the new era. Please feel welcome to follow up the 5G news and participate in discussions on my blog at:

▶ *www.5g-simplified.com*

I want to thank my colleagues and friends for providing me with such a great feedback and suggestions on the contents. I also thank my wife Celia and all my family for their support and patience when I worked on this guide.

I wish you a happy journey in yet another mobile communications era!

Jyrki Penttinen
Atlanta, GA, USA

What 5G is?

It is said that the telecommunications infrastructure is the biggest machine in the world. After the long-lasting era of the analogue switches interconnecting the phone users around the globe, the telecommunications systems have developed vastly especially in the past 20 years, connecting now people and *things*.

As soon as the first Internet Protocol -based data services were integrated to the mobile communication networks in early 2000s, there has been no way back to the limited platforms of the "old world". As an example, the home phones are about to go distinct thanks to the popularity of the advanced mobile communications, and wireless services form a truly integral part of our daily life.

We already take for granted the wireless multimedia devices and services. It is actually hard to believe how we managed communicating in older era, tight to the fixed location of the home phone—only couple of decades ago. So, let's start enjoying yet another mobile generation!

Generations

5G, or fifth generation, is the latest addition to the family of the mobile communication systems. It is a logical continuum to the previous generations that are categorized by different eras based on their technical capabilities. As a rule of thumb, each new generation provides significantly better performance compared to the previous ones. They also require new user devices.

The first generation, **1G**, became reality along with the commercialization of the first analogue mobile phones and networks throughout the 1980s. 1G referred to simple yet fully automatized mobile communications systems, and it was primarily designed for making wireless voice calls while on the move. In the 1G era, there were all kind of networks that typically did not work together. This lack of interoperability meant that the network services did not function while the customers were roaming abroad.

Nevertheless, 1G opened up the wireless era which turned out to be so popular that the new mobile network operators (MNO) had hard time forecasting the growth of the customer base. That caused MNOs frequent challenges in their efforts to timely construct the first cellular infrastructure and to satisfy the vast demand for the capacity and coverage. Despite somewhat light international roaming of some systems such as the NMT-900

(Nordic Mobile Phone), none of the 1G networks were truly global. By now, the 1G is practically obsolete as a result of much more efficient and performant new systems which took over the analogue era.

2G came into the picture in the early 1990s. Unlike the previous systems, the 2G systems are fully digital which enhanced the quality and brought along new services such as short messages and data transfer. Again, there have been many variants belonging to  2G. One of the most popular, GSM (Global System for Mobile Communications), serves still today many customers all over the globe. It was originally developed in Europe but it turned out to be universal. Meanwhile, the American 2G variant, CDMA (Code Division Multiple Access), has been serving customers in the domestic and international markets together with some other digital systems.

Today, 2G, and especially GSM, is still quite popular although its customer base is declining. This is because we consumers have possibility to use advanced systems that provide important benefits such as much faster data speeds, more generous capacity, and capability to serve significantly higher number of simultaneously communicating customers. Despite the other systems taking over and the foreseen sunset of 2G (see Figure 1), it is still a useful solution in many markets and may keep serving also current and new IoT (Internet of Things) devices such as wireless home alarms and sensors for years to come.

3G was designed to deal with even more efficiently multimedia applications such as video downloading. 3G has evolved throughout its lifetime offering increasingly fast data speeds, and it also introduced enhanced security level. The origi-  nally European HSPA (High Speed Packet Access) was called initially UMTS (Universal Mobile Telecommunications System) while the commercial system in the US is still known as cdma2000. Both are capable of providing tens of megabits per second (Mb/s) today.

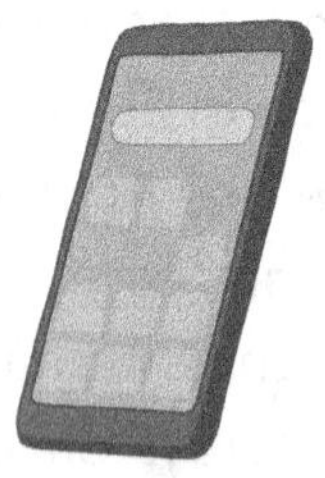 **4G** narrowed down the number of the systems that represent the new era. Today, the LTE (Long Term Evolution) is the only commercial 4G system after the relatively short market presence of its alternative, WiMAX2. LTE enhances the data speeds further and has brought along a renewed network architecture model which simplifies the system. LTE can handle hundreds of Mb/s data speeds at present.

5G is the newest generation, and it relies on a single, global standard. Its radio network is called New Radio (NR), and the fixed segment of the system, core network, is called Next Generation Core  (NGC). Together these two segments form a completely new 5G System (5GS). Even the 5G networks are in their infancy still in the beginning of 2020, they can already provide much higher data speeds compared to the LTE networks, and they exceed oftentimes several hundred Megabits/second (Mb/s).

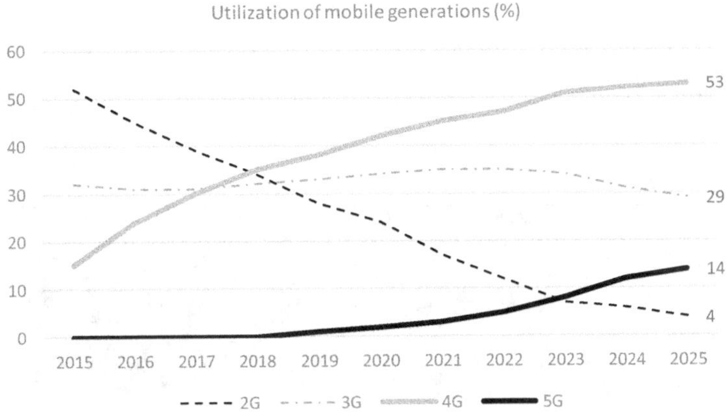

Figure 1 The statistics and foreseen development of the share of the current generations. Source: GSMA, The Mobile Economy 2018. (GSMA, 2019)

Statistics

One of the most popular services—after the voice-only 1G took off—has been the data transfer. It continues being the base for a growing number of services such as multimedia messaging and video streaming.

Figure 2 summarizes the development of the data services that different mobile generations are capable of supporting. As can be seen, the original data rates have been rather modest upon their first launch, and have increased quite considerably ever since as a consequence of the further development of each system. The estimate shows that each generation has offered us in average 10 times faster data speeds compared to the previous one.

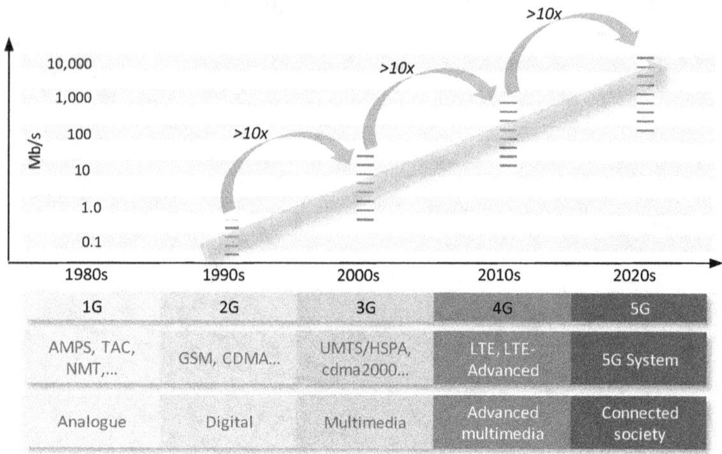

Figure 2 The main representatives and characteristics of different mobile communication generations.

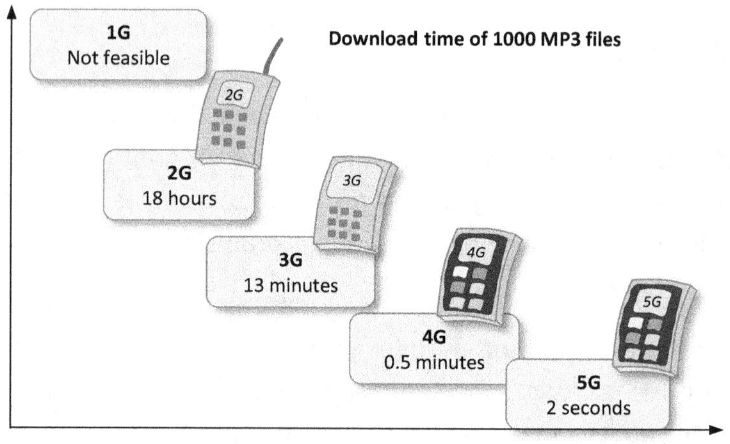

Figure 3 A rough comparison of the data performance of different generations in terms of download time.

Figure 1 shows the current and forecasted customer bases. 5G will enhance yet again the performance; not only in terms of the data speeds, but it also provides us users with some important additional benefits such as considerably lower response times (latency), and better reliability (lower down-time of the service). The security of the services has been upgraded, too, and the network architecture is completely renewed basing now on virtualized network functions. As a result, the 5G systems will outperform the previous generations in every aspect.

Today, compared to the original 9.6 kb/s, the evolved version of the 2G GSM is capable of handling download data speeds of up to roughly 500 kb/s, 3G HSPA+ up to 40 Mb/s, 4G LTE-Advanced up to 1 Gb/s, and 5G (theoretically as per Release 16) up to 20 Gb/s. Figure 3 summarizes the estimated transfer time for one thousand typical MP3 files, each 4 Mbytes in size.[1]

What Makes the 5G "Real"?

So, yet another generation has born. The first 5G system specifications have been ready since 2019, and the respective, fully interoperable and international standards-based networks have been launched soon after. The standardization body taking care of the 5G development is 3GPP, 3rd Generation Partnership Project, which also

---

[1] The bytes (B) are converted in this example to bits (b) by multiplying a single byte by 8 bits. 4 Megabytes (MB) thus equals to 32 Megabits (Mb).

keeps evolving the previous 2G (GSM), 3G (UMTS/HSPA) and 4G (LTE) specifications in a parallel fashion.

There were early 5G adopters in the markets already before 2019, well prior to the 3GPP-compliant 5G networks. Although those services were considered proprietary, they indicated how the then-forthcoming "real" 5G networks will look like.

What then makes the 5G networks "real"? The short answer is the compliance with the global standards, because they guarantee that the networks perform similarly everywhere and the users can roam with their compatible 5G devices any network that has roaming agreement with the user's own operator.

The global 5G requirements are presented in a concept called IMT-2020, International Mobile Telecommunications for 2020 and beyond. The ITU, International Telecommunications Union, which is the global authority of the telecommunications systems, has presented those requirements. The IMT-2020 summarizes a set of technical items and their values which dictate how capable 5G is assumed to be—at least in theory. (ITU, 2015)

While the systems may comply with these highly demanding requirements in ideal conditions, the practical performance is oftentimes somewhat limited depending on many realities such as the lack of radio coverage, the high number of simultaneous users sharing the network capacity, and the device's limited processing power, to mention some.

In any case, every new mobile communication generation has always enhanced significantly both the theoretical and practical capabilities, and so it is for 5G, too.

Some of the most important characteristics of the IMT-2020 include the capability to support theoretical data speeds of up to 20 Gigabits per second (Gb/s) for a single (only) user receiving contents (downlink) whereas the opposite direction, user sending data in uplink, is defined to support up to 10 Gb/s. Despite the fact that these values are challenging to reach in practice, at least in the beginning of the 5G deployments, they indicate the desire of the industry to offer us consumers much more demanding data services than ever before.

The 5G data speeds outperform easily those of the previous generations which is beneficial in, say, downloading large files on-the-go taken that the operator has implemented sufficiently 5G capacity within those areas. One example could be a video file downloading into the 5G device at an airport center just before the flight takes off. On the other hand, the typically larger bandwidth of the 5G cells means that the new system can support a greater number of simultaneously communicating high-speed data users.

The low latency is of utmost importance in 5G. It refers to the delay between the user's action with the 5G device, like a command to initiate file download, and the respective response such as the actual start of the download (see Figure 4). The ITU requires this latency value to be, in certain extreme use cases, as low as 1 millisecond (ms) which is considerably faster than any of the previous mobile networks could handle. That value is sufficiently good for critical and virtual reality applications.

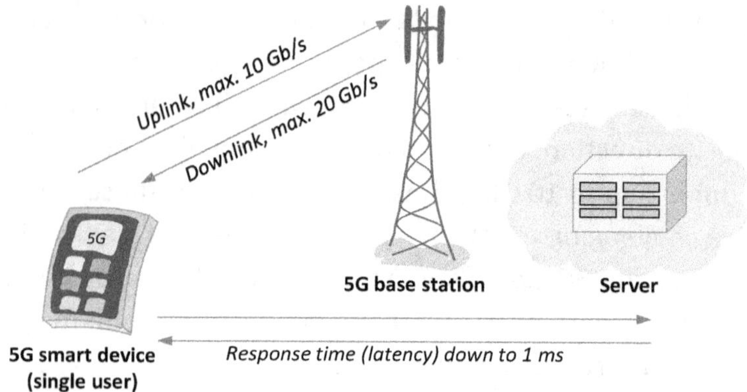

Figure 4 The enhanced data transfer from the 5G smart device takes place on uplink channels while the device receives data from the 5G base station on the downlink channels. The latency is better in 5G, too. Practical example is a video streaming; as soon as the user selects the video, the network starts delivering the contents from the video server almost immediately.

Thanks to such a low latency, 5G will provide us consumers highly demanding services such as real-time virtual reality games which are not necessarily feasible to use in previous cellular networks. The latency of the LTE has not been quite low enough for enjoying the most fluent virtual reality experiences so 5G provides finally a useful platform for those ultra-low latency environments.

The third new requirement relates to the reliability of the 5G services which can be up to 99.999%. This is an extremely demanding value in any environment, and it means that the cellular service should be up and running

99.999% of the time when the network is set to serve the respective ultra-reliable applications. Converting this value to an allowed "down-time" of the service, it equals to a maximum of half a minute of service outage during the whole month. As a comparison, if the reliability value is set to 99%, the allowed down-time can be up to about 7 hours per month. The ultra-reliable mode of 5G can serve real-time critical communications of which vehicle-to-vehicle messaging is one of the extreme examples.

The fourth requirement is specifically tailored for the massive Machine Type Communications (mMTC). It dictates that the 5G network should support up to 1 million simultaneously communicating devices per square kilometer (km^2). In this way, it would be possible to serve a huge number of machines and sensor devices on the field which is highly beneficial, say, for making the Smart City concept reality. Those sensors may include intelligent devices which measure the environment's data such as traffic flows, pre-digest the information and communicate the data to the servers that, in turn, combine the data from many other sources. 5G can thus form an important platform to serve the purposes of Big Data which refers to all kind of information the sensors and other sources measure, collect, correlate and analyze from the environment to facilitate the conclusions, find behavioral patterns, and help making decisions.

Nevertheless, these requirement values are applicable to only some of the use cases. It would not make sense to build up networks complying with all of these above-mentioned, highly demanding requirements permanently everywhere. That's why 5G is designed so that it can be adjusted to serve different user profiles upon need. That is something the previous networks have not been able to do as they provide same service level for all.

Building Blocks of 5G

As depicted in Figure 5, the basic pillars, or building blocks of 5G, are the evolved Mobile Broadband (eMBB), massive Machine Type Communications (mMTC), and Ultra Reliable Low Latency Communications (URLLC). These pillars are designed to serve optimally different user segments, or verticals, upon their need.

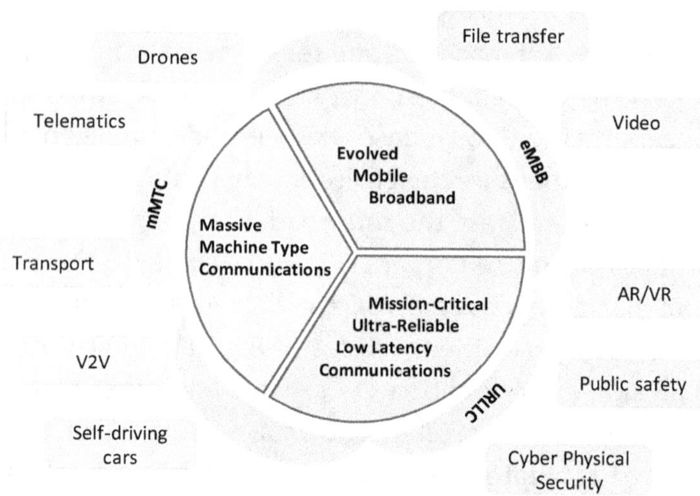

Figure 5 The main pillars of 5G are eMBB, mMTC and URLLC. They serve variety of users and verticals such as utilities, transport, telematics, critical infrastructure and public safety helping in their special communication needs.

▶Mobile Broadband

One of the 5G verticals is consumer market which benefits greatly from the high data speeds and capacity of the eMBB. We users belong to that category by default, and it provides us with the possibility to enjoy ever-increasing data speeds as the 5G deployments advance. As for the user's point of view, the eMBB of 5G means increased data speeds giving the possibility to stream, download and upload data much faster than in 4G.

In the very beginning of the deployments, the peak 5G data speeds are well below the theoretical maximum values of 20 Gb/s for a single user (in non-congested network) on downlink; the average 5G data speeds of some hundreds of Mb/s may be more realistic in loaded network. By the end of 2019, some of the public announcements showed around 1-2 Gb/s download speeds for 5G which indicates already the big potential of 5G. (Holland, 2019) Along with the further upgraded 5G networks and more capable 5G mobile devices (as per the 3GPP Release 16), the data speeds will keep increasing.

Wherever 5G will be gradually available, its eMBB mode provides very high data speeds clearly exceeding the previous 4G performance. The eMBB mode is included in the 5G system since the very beginning, but as the networks mature, the eMBB mode paves the way for much more demanding mobile broadband use cases such as AR/VR applications which may need to transfer data occasionally perhaps up to several gigabits per second.

▶Massive Communications

The massive Machine Type Communications relates to the Internet of Things (IoT) environment. It may include countless sensors on the field—that might communicate either actively or only rarely—and other types of communications between machines such as automatic car-to-car messaging to prevent accidents. Furthermore, many of the sensors may rely on an internal battery which needs to function on remote areas extremely long time without human intervention, perhaps up to 10 years. The power optimization is thus an important aspect of such devices.

The term mMTC means that 5G is able to support a huge amount of simultaneously communicating IoT devices whether they are for video transfer or for telematics. Its light version is included in the very initial 5G networks as a supporting service by the LTE system, and the next phase of 5G, based on the 3GPP Release 16 specifications, will enhance the respective performance further.

▶Reliable and Fast Response

The URLLC would be used in some very special environments, for example to provide the communication link within a critical infrastructure. It refers to a very reliable communication with extremely low delays.

In the most demanding case, the URLLC provides sufficiently good link for even remote surgery as a one of the potential use cases of 5G. In fact, some doctors in China have used 5G to perform remote surgery already in 2019. (Frost, 2019)

The URLLC provides thus fast responses and serves well for fulfilling the needs of the most critical communications where the reliability is important. Some examples of such cases include a drone flight control, critical factory automation, and wide-area smart city functions.

The Release 16 -compliant 5G networks will bring this mode into reality later in second half of 2020.

Networks Over Networks

5G has highly advanced means to serve the above-mentioned use cases by giving the operators possibility to form many types of networks in a parallel fashion within certain areas for different users. This functionality is referred to as network slicing which offers a network type that best serves the communications needs of its users such as fast data or low latency.

The network slices can be generalized as a "networks over network" concept. It means that the operator can adjust the performance of each network slice so that the users can benefit from the performance of the most ideal slice for their needs on the very same geographical area. Figure 6 clarifies the idea.

As soon as the operators will have deployed network slicing and other advanced functions, the 5G networks start offering personalized performance compared to any of the previous cellular networks that only provide one uniform performance type to be shared among all the users.

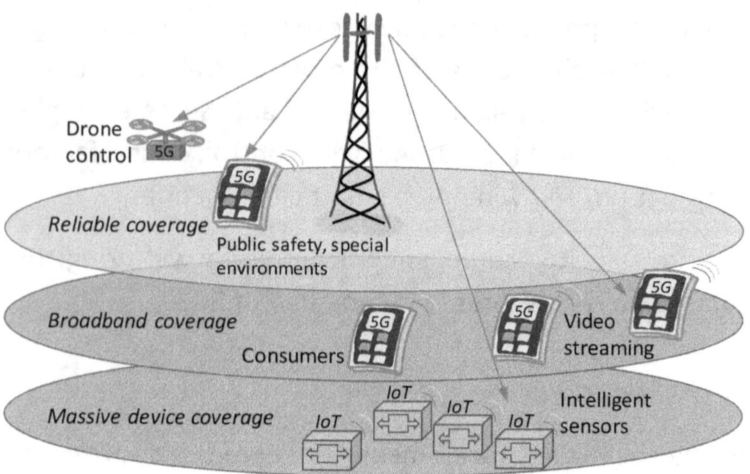

Figure 6 By adjusting different coverage types (network slices), the 5G operators can provide optimal performance for each user within the same area.

Development of 5G

Table 1 summarizes the key performance figures of 5G compared to 4G. Nevertheless, it is important to understand that the 5G networks will evolve gradually. The very first deployments are based on the early version of the 5G specifications called Release 15.

The 3GPP has produced these specifications in order to provide the operators means to start deploying the new networks in a fast pace in their initial, still rather limited form while the next specifications are under development.

The 5G operators will still largely rely on 4G infrastructure in the first phase throughout 2019-2020. This combination of the generations enables increased performance of the 5G system although it does not yet comply with the strictest requirements of the ITU's IMT-2020. (ITU, 2015)

The Release 15 networks will give us possibility to enjoy increased 5G data speeds of the basic services such as data file downloading and video streaming in limited areas. The operators and device manufacturers are currently working on the deployment of the "real", fully capable Release 16 -compatible 5G networks.

Table 1 Comparison of the 4G and 5G systems[2].

Performance	4G	5G
Data speed, peak value	1 Gb/s downlink (in theory)	20 Gb/s downlink (in theory)
User experienced data speed	10 Mb/s downlink (min. in practice)	100 Mb/s downlink (min. in practice)
Latency, min.	10 ms	1 ms
Device speed	350 km/h	500 km/h
Energy efficiency	1 x (reference)	100 x

Standardization

As mentioned previously, the ITU has an elemental role in defining how 5G will look like, and what kind of performance we consumers will start experiencing as soon as the ITU-compliant 5G networks are available for us.

The 3GPP is making the concrete technical specifications of 5G ensuring they fulfil the demanding requirements of the ITU. These requirements are listed in the ITU IMT-

---

[2] The theoretical 4G performance based on the ITU IMT-Advanced requirements, and 5G performance dictated by the ITU IMT-2020.

2020 vision, which can be found in their recommendation REC-M.2083. (ITU, 2015) The IMT refers to the International Mobile Telecommunications systems which are designed to comply with the needs of various use cases.

The ITU drives for advanced wireless infrastructure to connect the world, understanding that the broadband connectivity would be equally important as access to electricity. The IMT requirement set thus acts as a key pillar to enable mobile service delivery.

The ITU also foresees that in the future, both private and professional users will benefit from a wide variety of applications and services of the IMT systems. These include infotainment services and completely new industrial and professional applications. This development will have a positive impact on the integrated Information and Communication Technology (ICT) industry which will constitute a driver for economies around the globe. 5G is in a key position to make this evolution happen, increasing the economics and quality of life in a global scale.

The global principles of the radio frequency allocations are discussed at the Word Radiocommunication Conference, WRC. It takes place about every four years and it gathers the authorities and other involved parties to discuss the need and universal principles of the frequencies. WRC is organized by the ITU. The WRC-19 agreed about the allocation of further 5G bands like 26, 40 and 66 GHz.

How to Select the 5G Device?

In order to take advantage of the upgraded performance of 5G, one needs to obtain a new 5G cellular device—which is typically a smart device in the consumer market. Apart from the "traditional" and advanced mobile phone forms such as foldable displays, we can expect to see increasingly diverse 5G devices such as USB dongles, integrated 5G modules for connected laptops and home appliances, 5G-connected intelligent IoT sensors, advanced sport trackers and health devices, and embedded 5G devices for vehicles.

Challenges

The 5G system is still in its infancy in 2019-2020 which means there are inevitable coverage outages, and the 5G frequency bands are fragmented while the industry is advancing with the technology. During this time, we may expect to find 5G services mainly within certain city center areas, and the respective data speeds are not yet too fast compared to the theoretical maximum values.

Also, the first mobile device models might include merely basic 5G features and options such as limited support of the new frequency bands. One might thus wonder what the perfect timing to invest in the new phone would be.

The question is comparable to the purchase of any other new device such as laptop. Waiting for a bit longer, we might find more attractive models with advanced features, perhaps even at lower price. This is the case with the 5G mobile devices, too, so the decision is up to our personal preferences whether one wants to be an early adopter of enhanced yet basic features or prefers to wait for the more mature devices (and networks).

Regardless of the limitations of the first 5G devices, they are highly performant. Typically, they support multisystem communications so if there is lack of 5G coverage, the new phones can still serve within those areas.

Selection of the Model

There are already some 5G consumer devices available in the initial markets. If you consider to obtain a 5G-capable smart phone, the device manufacturers tend to provide information about the overall features, capabilities and generic performance values on their web pages.

Table 2 lists a snapshot of some of the available 5G-connected smart devices in January 2020 based on web.

Table 2 Examples of some 5G-capable smart devices in January 2020.

Maker	Model	5G bands (frequencies in MHz)	source
Honor	V30 / V30 Pro	1(2100), 3(1800), 41(2500), 77(3700), 78(3500), 79(4700)	GSMArena[3]
Huawei	Mate X	41(2500), 77(3700), 78(3500), 79(4700)	Huawei[4]
Huawei	Mate 30 Pro 5G	1(2100), 3(1800), 28(700), 77(3700), 78(3500), 79(4700)	Huawei
LG	V50 ThinQ 5G	41(2500), 260(39 GHz), 261(28 GHz)	LG[5]
Motorola	Moto Z2, Z3, Z4 Force, with moto5G module	Module designed for Verizon bands, see Motorola's web address[6]	Motorola[6]
OnePlus	7 Pro 5G	41(2500)	Oneplus[7]
Oneplus	5G McLaren	71(600)	Oneplus
Oppo	Reno 5G	78(3500)	Oppo[8]
Oppo	Reno3 Pro 5G	41(2500), 78(3500)	Oppo
Samsung	Galaxy Note10 5G	78(3500)	Samsung[9]
Samsung	Galaxy S10 5G	78(3500)	Samsung
Xiaomi	Mi Mix 3 5G and	78(3500)	Xiaomi[10]
Xiaomi	Mi 9 Pro 5G	41(2500), 78(3500), 79(4700)	Xiaomi
ZTE	Axon 10 Pro 5G	41(2500), 78(3500)	ZTE[11]

---

[3] *https://www.gsmarena.com/honor_v30_hands_on-review-2041.php*

[4] *https://consumer.huawei.com/en/phones/mate30-5g/*

[5] *https://www.lg.com/us/mobile-phones/v50-thinq-5g*

[6] *https://www.motorola.com/us/products/moto-mods/moto-5g*

[7] *https://www.oneplus.com/7t-pro-mclaren*

[8] *https://www.oppo.com/en/smartphone-reno-5g/*

[9] *https://www.samsung.com/us/mobile/galaxy-s10-5g/*

[10] *https://www.mi.com/global/mix-3-5g*

[11] *https://www.ztedevices.com/en/product/smartphone/axon_10_pro*

▶The Importance of Frequency Bands

The specifications define many 5G bands; the number is almost overwhelming. The positive side of this situation is that the operators can have more options to deploy capacity. The great challenge for the users is to find a device model that supports (sufficiently well) the selected operator's bands because they need to map.

We can have a look at the current devices' support of 5G bands in Table 2. First of all, every device of this table supports the 5G frequencies on the mid-band (1,000-6,000 MHz) with an exception of ONEPLUS 5G McLaren which has only one 5G frequency within the low-band region.

As will be explained later in the Frequency Bands section, the low-bad frequencies propagate farthest but have limited capacity (data speed), while the mid-bands offer a sweet-spot for the balance of the coverage and data speeds.

The fastest data rates can be obtained on the high-bands (6,000 MHz and beyond), but there is only one model in Table 2 equipped with high-bands; this device is LG V50 ThinQ 5G which also has one mid-band, 2,500 MHz (n41).

What is then the most adequate frequency band set that the 5G device would ideally support for the widest coverage? That is most likely the trickiest question of this book because there are overwhelmingly many options within each band type as indicated in Figure 7 and Figure 8, but the device manufacturers tend to implement only small part of them, and also the operators may have only few.

When you are considering a 5G device, the task is thus to compare its supported 5G bands with the ones of the operator of your preference and ensure that there is at least one common 5G band, or preferably more.

Let's have a look at some of the 5G operators to understand their deployed bands (this is not straightforward as the operator may have different regional frequencies depending on their licenses and deployment strategies).

One of the publicly available locations we may consult is the Wikipedia. Please note that even the correctness or freshness of this sort of source cannot be guaranteed, it gives a rough idea that can be confirmed directly with the operator you are considering. Table 3 summarizes the commercially deployed networks and their bands.

An important note for Table 3 is that there are many trials, too, which are commercialized soon. Nevertheless, how the operators exactly define the term "commercial" is up to their marketing but it typically means limited initial service areas while the operator keeps expanding their coverage area.

As we have learned from the previous generations, the roll-out of a completely new generation is a continuous work. Depending on the environment, it may take several years for an operator to reach, say, 90-95 % population coverage (meaning the people can have access to the 5G radio on those locations where they normally are), or 50% of the geographic coverage area of the country. As an example, Verizon announced their commercial mobile 5G launch in limited markets in Chicago and Minneapolis in 2019, and the work continues. (Comms Update, 2019)

Table 3 Examples of the commercial 5G frequency support of selected operators in January 2020. The bandwidth is in MHz. Interpreted from Wikipedia (https://en.wikipedia.org/wiki/List_of_5G_NR_networks).

MNO	Low-band n5, n28, n71		Mid-band n7, n38, n41; n78 (widest)		High-band n257, n258, n260, n261	
	Band	Width	Band	Width	Band	Width
Australia						
Optus			n78	60		
Telstra			n78	60		
Austria						
Drei			n78	100		
Magneta			n78	110		
Bahrain						
Batelco			n78	90		
stc			n78	90		
China						
China Mobile			n41	100		
China Telecom			n78	100		
China Unicom			n78	100		
Finland						
Elisa			n78	130		
Telia			n78	130		
Germany						
Telekom			n78	90		
Vodafone			n78	90		
Hungary						
Vodafone			n78	60		
Ireland						
eir			n78	80		
Vodafone			n78	100		
Italy						
TIM			n78	80	n258	200
Vodafone			n78	80		
Japan						
NTT Docomo			n78	100		
			n79	100		

Kuwait						
Ooredoo			n78	100		
stc			n78	100		
Zain			n78	100		
Latvia						
Tele2			n78	50		
Netherlands						
T-Mobile	n28	10				
New Zealand						
Spark			n7	20		
Vodafone			n78	50		
Romania						
Digi			n78	50		
Orange			n78	115		
Vodafone			n78	40		
Saudi Arabia						
stc			n41	90		
Zain			n78	100		
South Korea						
LG U+			n78	80	n257	800
KT			n78	100	n257	800
ST Telekom			n78	100	n257	800
Spain						
Vodafone			n78	90		
Switzerland						
Sunrise			n78	100		
			n80	20		
Swisscom			n78	120		
UK						
3			n78	100		
EE			n78	40		
O2			n78	40		
Vodafone			n78	50		
USA						
AT&T	n5	10			n260	100
Sprint			n41	40-60		
T-Mobile	n71	20			n261	100
Verizon					n261	400
Uruguay						
ANTEL (FWA)					n261	800

Equally important aspect is the operator's supported bandwidth per their each deployed 5G band. The wider the bandwidth, the higher the potential data speed. There is a maximum possible bandwidth per each band dictated by the specifications. Normally, each operator has only portion of these bands which can result in lower data rates depending on the demand on those locations.

It can be seen from Figure 7 that the low-band and mid-band typically only offer limited bandwidths even in the best cases, often only around ten or some tens of MHz. The exceptional mid-bands are n77, n78 and n79 which are located a bit higher on the spectrum, roughly within 3,500-5,000 MHz region. They provide much greater capacity via few hundreds up to near 1,000 MHz bandwidth values which, divided between the regional operators, serve rather well the capacity-hungry 5G services.

Observing Table 3, these specific bands seem to be especially popular in practical deployments thanks to their bandwidth benefits. As the mid-band frequencies have still rather decent radio propagation characteristics, they provide easily few kilometers radius for the urban and sub-urban cells offering rather high-speed data and feasible tecno-economic benefits for the operators.

Figure 8 reveals the real benefit of high-bands. They do not lack of bandwidth so depending on the operator's possibility to purchase capacity from these bands, they are capable of providing the highest data speeds in the form of small cells. Combined with low-latency edge computing, the high-bands offer adequate performance for the complex virtual reality services which are challenging—if not impossible—for low and mid-bands to handle.

Observing Table 3, the leader of the high-band deployments is South Korea with an 800 MHz bandwidth allocated to each one of the country's three operators. With this deployment strategy, after the commercial launch in the beginning of 2019, there were about 5 million 5G mobile customers by the end of 2019 making South Korea one of the most advanced 5G regions in the world at that time. (Mu-Hyun, 2019)

Also, Europe and the USA are advancing firmly with the commercial launches. As an example of the interest, the FCC has formed a fast plan for expediting the 5G deployments in the USA. The GSMA estimates that by 2025, 50 % of the population of the USA is covered by 5G while the global value is 15 %. (GSMA, 2019)

▶Advices for Selecting the Device

It is important to ensure the common bands of the operator of your preference and the 5G device you are considering, but there are also other aspects to be evaluated. It is recommendable to search such 5G-specific information about the device from the vendor's technical data sheets or public resources such as ***www.gsmarena.com***.

Table 4 collects some of the relevant 5G-related aspects you might want to consider when comparing different 5G smart devices.

Table 4 Some of the 5G-specific features and performance values.

Feature	Typical values
Bands	Low, mid and/or high-band (as per Table 2, Table 3, Figure 7, and Figure 8). Currently, the support of multiple 5G-specific bands on a single device is still rather limited but it can be assumed that there will be more models supporting multiple 5G bands rather soon.
Data speed	The theoretical maximum data rate of 5G could be up to 20 GHz (downlink) and 10 GHz (uplink), and few hundreds of Mb/s to few Gb/s in practice for multi-user coverage area. The data sheet of the device indicates the maximum value the device is capable of supporting in the best case, referring to the supported data capabilities of the modem chipset.
MIMO	Multiple In, Multiple Out antenna. In 5G, this value can be anything between 2x2 and 8x8 (if there is no MIMO, it is actually 1x1 MIMO forming only one link between the transmitter and receiver). The more the device has MIMO antenna elements, the faster data rates it provides—if only the network supports the same combination. The MIMO may also help in increasing the total received power (field strength). Today, many LTE devices support 2x2 MIMO, and there are already some high-end mobile devices supporting 4x4 MIMO. The grade of MIMO depends on the modem chipset of the device. As an example, Qualcomm Snapdragon Gigabit LTE supports it.[12]
CA	The CA of the LTE and 5G data refers to Carrier Aggregation. The higher the number, the faster data speeds the device can offer for LTE and 5G thanks to the feature's ability to use parallel bands simultaneously.
Modulation	5G has many modulation schemes which work automatically. The latest one, 256-QAM (Quadruple Amplitude Modulation) provides fastest data in reduced areas whereas the other schemes work farther.

---

[12] *https://www.qualcomm.com/snapdragon/gigabit*

It is unlike that we'll see solely 5G devices anytime soon due to the initial lack of 5G coverage area. The new smart phones support thus typically both 4G LTE and 5G systems, combined often with a 2G and/or 3G. Such device is capable of executing handovers between systems.

It is worth noting—even many operators are planning to switch off their 2G and/or 3G —that there may still be areas for years to come especially in economically limited countries which do not have LTE nor 5G.

As for specifically 5G connectivity, the most important features that make the difference between 5G and previous generations, and that are relevant for the comparison for different 5G models, are the supported bands (and their bandwidths, the widest bands providing fastest data speeds), antenna configurations (the more multiple elements, the higher data rates), and supported modulation schemes (highest grade provides best performance but only very close by the base station).

Furthermore, you might want to check whether the device supports at least one useful segment on the low-band (less than 1 GHz), mid-band (1-6 GHz) and high-band (more than 6 GHz). Table 4 presents some of the items.

If you are going to use the device primarily on the home network and purchase a respective operator-branded device, it is probably already tailored to support the respective frequency bands in an optimal way. Nevertheless, for the international use, it is worth checking the supported band list apart from the home operator's own frequencies to ensure the best coverage in the visited regions.

As can be seen from the summary presented in Table 2, the current 5G devices typically support only a small fraction of the bands of Figure 7 and Figure 8.

Furthermore, as can be interpreted from Table 3, the operators typically have significantly smaller share of the theoretical frequency bands. This is because the operators bid and purchase blocks of the bands, each winner getting their own share of the spectrum and operating it adjacently with other competitors in the area.

This means that even if the 5G device is capable of transmitting on, say, 100 MHz channel of the band that the operator supports, and if the operator has a license to operate only a 20 MHz block of that specific band, the achievable data speeds may be considerably lower than could be achieved in theory.

In addition, the operators may have license for operating some of these frequency bands and their blocks only regionally, so the number of bands and their bandwidths, and thus the data speed may vary between the locations of a single country.

The spectrum authorities are in a key position to facilitate the deployments of the operator community. As an example, the FCC regulates the 5G (and other commercial) bands in the USA.

Continuing the same example, the FCC is driving for additional spectrum available for 5G services on low, mid and high-bands, indicating the following: (FCC, 2020)

✓ **High-band**: The FCC makes available 24 GHz and 28 GHz bands, as well as 37 GHz, 39 GHz, and 47 GHz bands, corresponding about 5 GHz of 5G spectrum.

✓ **Mid-band**: The FCC is making available 2.5 GHz, 3.5 GHz, and 3.7-4.2 GHz bands, corresponding 844 MHz for 5G deployments.

✓ **Low-band**: The FCC is planning to make available 600 MHz, 800 MHz, and 900 MHz bands.

✓ **Unlicensed bands**: The FCC is planning to make spectrum available also for next generation of Wi-Fi in the 6 GHz and above 95 GHz band.

Each band is typically divided between multiple operators, and oftentimes, an individual operator cannot offer the full theoretical capacity for their customers at least in the beginning of the new networks.

As the commercial markets evolve constantly, it is thus recommendable to:

✓ Ask from the mobile operator's customer service that they have a license to offer sufficiently capacity (data speeds) within the area of your interest;

✓ Ensure that the 5G smart device supports the compatible bands and sufficiently wide bandwidths.

▶ Non-Cellular Features

The non-cellular-related data of the device include the display size (diagonal, inches of screen-to-body ratio) and form (typically full display with virtual keyboard), internal storage capacity (Gigabytes), and camera type and its resolution (pixels, optical zoom, image stabilization, aperture, sensor size, number of cameras).

The manufacturers also indicate the battery capacity (mAh), charging characteristics (minutes to full and partial charge, wireless or via connector), and additional accessories such as fingerprint reader, Artificial Intelligence (AI) engine, and supported connectivity types.

Oftentimes, the vendor sources do not emphasize the actual 5G-related features too much, and only in generic terms such as the processor type (e.g., Qualcomm Snapdragon 855) but not necessarily detailing the actual activated set of features of the processor or device.

This is the most challenging part when comparing the performance of the 5G models. The 5G-capable devices tend to have high performance capabilities and include advanced functions of high-end category by default. Table 5 presents the typical features to ease the comparison. The weight of each item depends on each one's personal needs.

The added value pf the 5G device derives from the technological features such as more performant antenna types (multiple in, multiple out, or MIMO), wider frequency bandwidths, and the number of supported bands.

Table 5 Some of the typical technical items and their values the 5G devices support in a 5G-agnostic way.

Item	Typical values
Multi-system	Support of 5G, 4G, possibly 3G, 2G; helps to ensure service continuity also in 5G outage areas.
Price	There will be an increasing amount of different types of 5G devices from dedicated sensors (sub-10 usd), embedded modules (sub-100 usd) and the most basic phones (sub-300 usd) up to the most complex smart devices (1500+ usd).
Screen size	Screen size follows the established LTE device principles, being oftentimes in the range of 4.5-6.7 inches.
Battery capacity	Typically, 2000-4500 mAh; the offered battery capacity indicates roughly the usage time per single charging for light, medium and heavy use depending on the model's variable features, but it depends on the processor's capabilities, too. The most performant chipsets consume more battery and require typically 3000 mAh or more for decent duration, which increases the device size.
Idle mode duration	The battery typically lasts several days if the phone is not used. For sensors, the need may be even up to 10 years with a single charge.
Battery charging time	Similar with the LTE devices, e.g., some tens of minutes on fast charging, and 30-90 min on normal charging. Can differ on wireless charging.
ROM, internal	Internal storage for file system and OS. 32-256 GB, nowadays also 512 GB or 1 Terabyte (TB).
RAM, internal	Stores critical files needed by the processor, e.g., OS components and application data. Often, 4-12 GB. The larger RAM the more fluent the usage.
Memory card support	As has been the case in LTE, some models support typically 512 GB or 1 TB memory cards while there will be also models not supporting the card.
Camera	Similar as in LTE, e.g., 8-16 Megapixels (Mpix), sometimes 20-48 Mpix.

Connectivity	Some or all of the following: Bluetooth (BT), Wi-Fi 802.11ax, Near Field Communications (NFC), satellite navigation (multi-system), USB.
Audio	3.5 mm line, BT, or vendor-specific connection.
Output power	Typically, the maximum defined by the 3GPP per band and mode (e.g., 23 dBm, or 0.2 Watts). For the fixed wireless access, the maximum power may be higher (35 dBm, or 3 W).

Optimizing the Cost

The purchase of a 5G device is about how to balance the costs, features and performance. We might want to set a maximum target price for the device and try to stick to the budget—knowing the utilization of the most advanced features may require a bit more expensive subscription (data plan) of the operator, too.

Let's investigate some of the available models and what they offer in each price category based on the summary of Table 2. Please note that this example represents only a snapshot of the global markets, and that there increasing number of new devices keep becoming available, but it serves to make similar exercises of your own.

Let's collect four models from Table 2, each representing different price points. Please note that these models are selected into this list randomly in order to have means to compare typical features per selected price range.

Table 6 presents this summary interpreted from the data presented at ***www.GSMArena.com*** on January 2020.

Table 6 Comparison of certain 5G devices. Data collected on January 2020.

Maker:	Honor	ZTE	Samsung	LG
Model:	V30	Axon 10 Pro 5G	Galaxy S10 5G	V50 ThinQ 5G
Price, usd	~485	~790	~840	~1200
Display, inch	6.57	6.47	6.7	6.4
Display, pixels	1080x2400	1080x2340	1440x3040	1080x2340
Display, density, ppi	400	398	502	403
Screen/body, ratio, %	84.9	87.9	89.4	83.3
Thickness, mm	8.9	7.9	7.9	8.4
Weight, g	213	175	198	192
Android	10.0	9.0	9.0	9.0
Battery, mAh	4200	4000	4500	4000
RAM, GB (max)	6	12	8	8
ROM, GB (max)	128	256	512	256
Memory card, TB (max)	N/A	1	N/A	1
Modem	HiSilicon Kirin 990	Qualcomm SM8150 Snapdragon 855	Qualcomm SM8150 Snapdragon 855, (USA); Exynos 9820 (Global)	Qualcomm SM8150 Snapdragon 855
Camera, /f	1.8, 2.4, 2.4	1.7, 2.4, 2.2	1.8, 2.4	1.8, 2.4
Camera resolution, MP	40, 8, 8 (triple)	48, 8, 20 (triple)	16, 16 (dual)	12, 13 (dual)
Audio, 3.5 mm jack	No	Yes	Yes	Yes
2G Bands, quad, GSM	Yes	Yes	Yes	Yes
2G Bands, dual, CDMA	No	Yes	Yes	Yes
3G bands, quad, HSPD	Yes	Yes	Yes	Yes
3G bands, cdma2000	No	No	Yes	Yes

4G bands, LTE global	1, 2, 3, 4, 5, 7, 8, 12, 17, 18, 19, 26, 34, 38, 39, 40, 41	1, 2, 3, 4, 5, 7, 8, 12, 13, 17, 18, 19, 20, 26, 28, 34, 38, 39, 40, 41	1, 2, 3, 4, 5, 7, 8, 12, 13, 17, 18, 19, 20, 25, 26, 28, 32, 38, 39, 40, 41, 66	unspecified
4G bands, LTE USA	unspecified	unspecified	2, 3, 4, 5, 7, 8, 12, 13, 18, 19, 20, 26, 28, 38, 39, 40, 41, 46, 48, 66	1, 2, 3, 4, 5, 7, 8, 12, 13, 17, 20, 25, 26, 28, 40, 41, 46, 66, 71
Bands, 5G	1(2100), 3(1800), 41(2500), 77(3700), 78(3500), 79(4700)	41(2500), 78(3500)	78 (3500)	41 (2500)
Cellular connectivity	GSM, HSPA, LTE, 5G	GSM, CDMA, HSPA, LTE, 5G	GSM, CDMA, HSPA, EVDO, LTE, 5G	GSM, CDMA, HSPA, EVDO, LTE, 5G
SIM connectivity	Dual SIM (Nano-SIM, dual stand-by)	Dual SIM (Nano-SIM, dual stand-by)	Single SIM (Nano-SIM) or Dual SIM (Nano-SIM, dual stand-by)	Single SIM (Nano-SIM) or Dual SIM (Nano-SIM, dual stand-by)
Data speeds	HSPA 42.2/5.76 Mbps, LTE-A, 5G (2+ Gbps DL)	HSPA 42.2 / 5.76 Mbps, LTE-A (3xCA) Cat18 1200 / 150 Mbps, 5G (2+ Gbps DL)	HSPA 42.2 / 5.76 Mbps, LTE-A (7xCA) Cat20 2000 / 150 Mbps; 5G (2+ Gbps DL)	HSPA 42.2 / 5.76 Mbps, LTE-A (7xCA) Cat20 2000 / 150 Mbps; 5G (2+ Gbps DL)
WLAN	Wi-Fi 802.11 a/b/g/n/ac, dual-band, Wi-Fi Direct, hotspot	Wi-Fi 802.11 a/b/g/n/ac, dual-band, Wi-Fi Direct, hotspot	Wi-Fi 802.11 a/b/g/n/ac/ax, dual-band, Wi-Fi Direct, hotspot	Wi-Fi 802.11 a/b/g/n/ac, dual-band, Wi-Fi Direct, DLNA, hotspot
Bluetooth	5.1, A2DP, LE	5.0, A2DP, LE	5.0, A2DP, LE, aptX	5.0, A2DP, LE, aptX HD

GPS	A-GPS, GLONASS, BDS, GALILEO, QZSS	A-GPS, GLONASS	A-GPS, GLONASS, BDS, GALILEO	A-GPS, GLONASS, GALILEO
NFC	Yes	Yes	Yes	Yes
FM radio	No	No	Yes	Yes
USB	2.0, Type-C 1.0	2.0, Type-C 1.0	3.1, Type-C 1.0	3.1, Type-C 1.0

As can be seen, the more expensive, the better equipped the device typically is—although not necessarily always and for every feature. It is interesting to note that none of the models presented in Table 6 have 5G high-bands; of the overall market snapshot, Table 2 indicates that the LG model is the only one supporting high-bands, or mm-Wave bands of 260 (39 GHz) and 261 (28 GHz).

The next step is to ensure the device has all the desired functions you might want to use in the foreseen near future. As an example, for the ones preferring to see video contents on-the-go, a device with sufficiently wide screen (6.7" or more) makes sense.

For the ones traveling a lot, the global roaming capability is important, meaning that the device should have various different bands—preferably at least one popular frequency from low, mid and high-band categories.

We also might want to evaluate the opportunities for the financing. Some operators offer down payment plans during certain time period, meaning that you can invest in the phone deferring the final payment with a low interest rate. Typically, these devices are locked on the respective operator's network until the payment is completed. The technical benefit of such devices is that the operator has most probably ensured with the manufacturer that the operator-specific requirements are supported and that they perform as expected.

Another option is to obtain an unlocked device which would work on any operator's network as long as the device is compatible with the used system and its radio capabilities (bands).

It should be noted, though, that not all the operators provide their full set of possible services if you bring your own device. One example of such limitation is the possibly disabled Voice over LTE (VoLTE) or Rich Communications Services (RSC) for the open devices, so it is a good idea to request information from the operator about the support of your preferred features, and whether there are any potential limitations, in order to avoid surprises after the contract has been signed off.

Features

The 5G devices support typically the same kind of features and functionalities as any other smart device. As the 5G networks evolve further, we can expect to see more advanced 5G services and devices with new features and better performance; this is a familiar pattern from all the previous generations.

The special aspects of the initial 5G devices are related to the faster data speeds, which may require more powerful processor. The sufficient processor performance depends largely on the application you are going to use. It can be assumed that the 5G devices will be a base for all kind of advanced services such as 3D/360-degree video and virtual/augmented reality applications which require a rather powerful device, large memory storage, and high-capacity battery.

These aspects indicate that the 5G device that would be capable of taking full advantage of the advanced services needs to be of the highest end category, and of 3GPP Release 16 compatible in the most advanced stage.

If you plan to use a 5G-capable device mainly for light, "traditional" applications, its key benefit might be merely the additional 5G radio connectivity for an alternative service coverage, the increased data speed being a secondary benefit so one option is to consider waiting for more economic 5G devices to enter the markets.

After the first stage of the 5G deployments, we may see 5G-capable basic models also in the sub-300 usd category. As one of the benefits of 5G is the high data speed per se, it requires rather capable processor that may not be available in the lowest price categories anytime soon.

Related to the 5G-agnostic aspects, such as camera and display characteristics, the following sections present guidelines for selecting the most appropriate device depending on your needs.

▶Display

The display size of the 5G device, or any other device, is a tradeoff of the usability—how well the device fits into the pocket—and the user experience. The larger display looks better for viewing video contents and photos. The display size of the mobile devices varies typically between 4.7–6.9 inches, new foldable, bigger displays entering markets, too. You might want to consider the pros and cons of these aspects taking into account the possibly increased price tag of the larger displays.

The number of the pixels dictate the resolution. The more pixels, the more accurately the device shows the images.

The pixel density indicates the quality of the picture and video contents. The typical minimum value for the mid-tier category of 300 pixels per inch (ppi) provides a good quality, but 400-500 ppi might be desirable for the more advanced video display. It is recommendable to test some visually critical applications before deciding to go for certain ppi value because the higher the density, the higher the cost and processor power requirements are.

The screen-to-body ratio relates merely to the aesthetics of the device; the less bezel (non-display area) round the sides, the more appealing the device might look like—depending completely on the personal taste.

▶Physical Aspects

The weight and thickness of the 5G device tend to be roughly comparable to the devices of similar categories.

To maintain the desired values (say, below 200 g and 8 mm), the device manufacturers need to balance the supported functionalities such as the additional features, number of bands, size of the battery etc. Typically, the small quality components cost more than the older, robust ones so the device maker needs to balance between many tecno-economic factors when designing the mobile device's price category. The quality has its price which typically is reflected in the cost for the consumer. Please note that the quality refers to the inner components and the materials of the body, display and accessories alike.

▶Operating System

The operating system (OS) of the 5G device is typically Android, or alternatively iOS as soon as Apple starts supporting 5G bands. The OS is normally upgradeable. That is not for granted for all the older models, so it is a good idea to ensure the future support with the manufacturer.

▶Battery

The 5G device requires more powerful processor and thus a higher-capacity battery as a consequence of the higher data speeds, although the future processors are constantly being optimized. The maximum battery capacity tends to be up to 4000-4500 mAh which has impact on the size and weight of the device. The power of the charger is important, too, to ensure it charges fast.

The duration of the device's battery per charge cycle depends on the applications and functions, and their utilization time. If the battery duration in normal utilization is less than one complete day per charge cycle, it may have highly negative impact on the user experience.

The smart device's battery loses gradually its capacity, so it needs to be replaced perhaps every 2-3 years. Today, nearly every model has in-built battery so the replacement requires expertise (certified maintenance service) unless you prefer considering any of the limited number of models housing changeable battery.

▶Memory

The modern apps, high-resolution photos and videos require increasing memory capacity. The 16 GB or 32 GB internal memory is getting nowadays challenging to use with such a big amount of data if we want to store on it.

Some models have a slot for external memory card (SSD, Solid State Drive) which may be typically of up to 1 Terabytes (about thousand Gigabytes). For the models not supporting external memory card, the contents can be uploaded onto a cloud. Some cloud providers may want to charge for the storage use, though. It is matter of personal taste if you prefer storing the contents locally on the memory card (maintaining complete control), or remotely on the cloud (beneficial if you lose the device).

▶Modem

The modem makes the cellular connectivity possible. There are only few 5G modem providers in the beginning of the 5G deployments, but the number increases later. Also, the modems will support more evolved data speeds.

Although many mobile devices rely on the same modem types, the device maker does not necessarily activate all of the features such as RF bands, as can be seen in Table 5. As a rule of thumb, the more 5G bands the device supports, the better compatibility it provides when changing operators and when roaming abroad, but the more complete band set may also increase the price of the device.

▶ Camera

The camera performance is important for the ones wanting to take high-quality photos. The manufacturers typically inform the f/ value. The smaller the number, shallower area of the focus. In practice, the small number provides sharp focal picture with blurred background which provides more professional looking pictures. Also, the speed of the camera application is important.

The 5G devices may support advanced camera features such as multiple camera lenses, image stabilization, and large sensor although the quality of the physical optics is still in a key role. The Optical Image Stabilizer (OIS) may improve the quality of the photo considerably but it can also increase the cost of the device.

Known and established camera manufacturer brand is probably a safe bet when deciding between different optical lens models although this is not necessarily to be taken for granted in such a dynamic business.

It is thus recommendable to test the camera performance before the purchase, experimenting scenarios such as daylight, night, and fast-moving objects. The personal preferences and the own interpretation of the quality beyond the informed values may vary a lot in practice.

▶Audio

The audio performance may differ vastly between the devices. Some models handle high-volume recording easily while others cause distortion and noise, so again, it is a good idea to test the recording and playback performance before the purchase.

As for the earbud connectivity, some manufacturers are moving away from the traditional 3.5 mm audio jack. The benefit of the jack is the wide availability of 3rd party earbuds while the new connectors may work with adapter, or rely on wireless connectivity such as Bluetooth.

In the latter case, check the support of high-quality stereo Bluetooth profile if the audio performance is important to you. One of the benefits of the wireless or proprietary audio connectivity may be their better water resistance.

▶Cellular Connectivity

Communications-wise, the most important functionality of the mobile device is the cellular connectivity, so the more systems and bands the device support the better. It is important to ensure the device's bands coincide at least partly with your preferred operator.

For the 5G connectivity, the first models have only few 5G bands (or only one 5G band) so it is import to ensure they are supported by your operator; otherwise the device would not be able to take advantage of the new 5G radio.

▶SIM

The 5G system still relies on the SIM card. It can be of traditional, removable card or an embedded module. The device may also be able to use the new eSIM concept basing on the removable or physically embedded SIM.

The device may support dual or multiple SIMs which the user can activate in a parallel fashion. The dual SIM enables the use of two networks at a time. If the device supports the dual SIM, it may offer some of the following options (please note there may be also other type of options available):

✓ Passive; the device may have two SIMs, but only one can be selected for use at a time. Passive Dual SIM device is in reality a single SIM equipment with a single transceiver and it forms a logical connection to the operator that is selected from the menu.

✓ Dual SIM Dual Standby (DSDS); both SIMs can have idle-mode network connection (waiting for the calls), but when the actual connection is activated, the second connection is disabled. The SIMs in this case share a single transceiver, and the registration to the second network is maintained in the idle mode and during the call.

✓ Dual SIM Dual Active (DSDA); this is the most compete mode as both SIMs can be used in both idle and connected modes. Each SIM uses their own transceiver. It is worth noting that some DSDA devices have only GSM for the second transceiver. (GSMA, 2018)

▶Data

Depending on the modem, the RF band provides certain maximum data speed. The practical data speed may not reach the theoretical value as the operator may lack the maximum capacity (bandwidth), and in practice, the capacity is shared amongst the other users. Nevertheless, the device's support of those theoretical speeds indicates the capabilities for the comparison of the 5G devices.

▶Protection, Solids and Liquids

Some devices have enhanced protection against solids and liquids. This is referred to as Ingress Protection (IP). As an example, 5G device complying IPx8 means it can be used under water (camera) but not for data transmission.

Table 7 The IP classification. The IP consists of 2 digits, IPab of which the "a" is the class for protection against solids whereas the "b" refers to the protection against liquids as per the IEC 60529 standard.

Class	a (solids)	b (liquids)
0	No protection	No protection
1	Objects > 50 mm	Vertically dripping water
2	Objects > 12.5 mm	Spraying water, up to 15 deg from vertical
3	Objects > 2.5 mm	Spraying water, up to 60 deg from vertical
4	Objects > 1 mm	Spraying water, all angles
5	Limited against dust	Water jet, any direction
6	Complete against dust	Powerful water jet, all
7	N/A	Immersion in water, up to 1 m depth, 30 min.
8	N/A	Immersion in water, over 1 m (with added statement)
9K	N/A	Water jet, hot/high pressure
X	Not tested / certified	Not tested / certified

Non-Cellular Connectivity

The 5G devices may support Wi-Fi, FM radio, USB and other non-cellular systems such as one or more positioning systems. Not all these are supported in every model so if you need any of these functions, please ensure the device has the latest versions. The following summarizes the key aspects of the connectivity options.

▶Wi-Fi

The 5G devices may support a number of Wi-Fi variants. It is recommendable to consider the latest standards such as the IEEE 802.11ax[13]. It refers to the Wi-Fi 6 which provides fasted data speeds up to date, about 10 Gb/s.

▶Bluetooth

The Bluetooth (BT[14]) standard keeps evolving, and the latest standard is currently the BT 5.2 (January 2020). The standard defines various optional profiles so it is worth checking that the cases you foresee to use are supported, such as high-quality stereophonic audio over BT.

▶Near Field Communications

Near Field Communications (NFC[15]) is the base for the current payment solutions such as Google Wallet and Apple Pay. It is also useful for small-scale data transfer between two devices (for business card exchange etc.), or

---

[13] *https://standards.ieee.org/project/802_11ax.html*
[14] *https://www.bluetooth.com/specifications/bluetooth-core-specification/*
[15] *https://nfc-forum.org/*

for pairing devices and activating functions such as Bluetooth link which can be established by placing first the NFC devices close to each other to trigger the BT.

▶Positioning Systems

The 5G devices typically support at least one positioning system which is by default GPS (Global Positioning System). The device may also support other ones such as Galileo (European), GLONASS (Russian), BDS (Chinese) or QSZZ (Japanese), forming a global navigation satellite system (GNSS[16]). The more the systems are supported, the better location service coverage can be achieved although that may have impact on the battery consumption.

▶USB

The data transfer can be done via the Universal Serial Bus (USB[17]). There are different versions available, and the 5G devices typically support the USB Type-C 2.0 or the more modern Type-C 3.1. The USB is also used for charging.

Some devices can support the OTG (On-the-Go) functionality which refers to the reversed cabling of the USB. So, for example, you can connect your memory stick to the OTG USB port and instead of only transferring data between the device and memory stick, you can have for example video streaming from the memory stick projected to the device's display.

---

[16] *https://gssc.esa.int/navipedia/index.php/An_intuitive_approach_to_the_GNSS_positioning*
[17] *https://www.usb.org/*

▶Video

The 5G devices support video output likely via wireless casting. The old mini or micro HDMI (High Definition Multimedia Interface) connector is not expected to be too typical in 5G devices. Instead, micro-USB port is more probable choice which you can connect via an USB-to-HDMI converter to the TV. The variants include Mobile High-Definition Link (MHL) and Slimport.

Special and IoT devices

For the special 5G devices, such as IoT sensors, connected autos and factories, drones, and law enforcement equipment, there are aspects that differ from the consumer markets. One of these is a direct connectivity between the 5G devices which can be highly useful for the rescue crew communications, especially in cellular outage areas.

The network and mobile device specifications leave many of the features optional so it is up to the device maker to integrate them whereas the operator can have a choice to activate or deactivate them. If you are interested in a specific 5G feature, it is thus a good idea to consult the operator and device manufacturer to ensure it is supported from both network's and device's side.

Processor and Modem

The foundation of the 5G device is the processor which executes the HW/SW functions, and the modem chip to handle the cellular radio connectivity. The performance of the device depends on their capabilities and features, together with the memory capacity. There are many 5G chipset makers in the market providing the processors and modems, and more is expected to onboard.

The 5G modem chips typically support multimode radios within the same physical System on Chip (SoC), so in addition to the 5G New Radio (NR), with its intermediate, 4G-assisted (Non-Standalone, NSA) mode, or the "real", 5G-only (Standalone, SA) mode, the same chip may integrate also 2G, 3G and 4G modems.

Table 8 Snapshot of current 5G modem manufacturers and the key performance figures of the chips. Comparison is based on public sources.[18, 19, 20]

Maker	5G modem	Key figures
Qualcomm	Snapdragon X50	Download speed up to 5 Gb/s; connected to a separate Snapdragon 855 processor and LTE modem.
Samsung	Exynos 5100	Download speed up to 6 Gb/s; integrated with 2G-5G radios.
Huawei	Balong 5G01	Download speed up to 2.3 Gb/s.
MediaTek	Helio M70	Download speed up to 4.7 Gb/s.
Intel	XMM 8060	Download speed up to 6 Gb/s; integrated with 2G-5G radios.

---

[18] *https://www.netscribes.com/5g-modem-chipset-manufacturers-smartphones/*
[19] *https://www.gsmarena.com/mediatek_announces_helio_m70_with_5g_modem-news-37280.php*
[20] *https://newsroom.intel.com/news/intel-accelerates-timing-intel-xmm-8160-5g-multimode-modem-support-broad-global-5g-rollouts/#gs.mkn93t*

The 5G chipset manufacturer may also provide the 5G modem as a separate component within the device, which may not be optimal as for the energy efficiency and physical real estate are considered. For that reason, it is assumed to be an intermediate solution as 5G matures.

Table 8 summarizes some of the current 5G modem manufacturers and their key products.

The 5G device ecosystem consists of these elements:

✓ 5G system (5G radio and 5G or 4G core depending on the use of the SA or NSA modes, respectively) which provides a set of 5G-specific functions; the initial capabilities will be enhanced and complemented later.

✓ 5G devices which consist of new processors and chips to support the 5G cellular connectivity; the modems and chips support the initial functions that do not provide yet the full performance of the most demanding requirements of the 5G as per the ITU IMT-2020.

The operators are building up the 5G ecosystem gradually, and as the networks start supporting the 3GPP Release 16 specifications and Standalone (end-to-end 5G system), the 5G performance increases significantly.

Meanwhile, we can compare the performance figures of the current 5G devices and decide whether their differences are important enough to prioritize any of the device over other, including memory, display size and its form (the foldable or dual displays becoming likely more popular), and battery charge duration.

Future Models

The evolution happens fast and new, enhanced models will be introduced to market regularly. Nevertheless, as can be seen from the presented download data speed values, the current 5G devices do not support the maximum ITU IMT-2020 data speeds of 20 Gb/s quite yet.

In the future, we can expect to see totally new type of 5G devices. One of these may be a Virtual Reality (VR) and Augmented Reality (AR) headset with an advanced 3D display and 360-degree dynamic viewing angle. These devices can be used in gaming and other environments.

Some of these devices may be able to offload the data processing to the 5G core network relying on the ultra-low latency and evolved mobile broadband data speeds. The offloaded processing could be feasible for enabling low-cost multimedia devices without high-performance processor, apart from the capabilities needed for displaying the post-processed contents from the cloud.

The 5G devices will diversify further as the 5G infrastructure will have evolved functions. The respective applications may lift medical imaging, advanced gaming and video conferencing into new level. As an example, the idea of the 5G-based 3D-holographs have already been demonstrated in practice by Ericsson and Vodafone.[21]

---

[21] *https://www.ericsson.com/en/news/2018/11/3d-holographic-calls-with-5g*

How to Select the 5G Operator?

Some of the mobile network operators have already deployed the initial 5G infrastructure and are providing 5G-specific wireless plans for the customers. Before using these 5G services, we customers need a new 5G mobile device and the respective subscription in order to start experimenting the enhanced performance.

Yet, there is only a handful of 5G-capable mobile devices available in the beginning of 2020. As their 5G-specific functions are still rather limited, we might want to ensure the features of the device and the network map with each other.

What We Need to Know?

Prior to selecting a new 5G contract, we need to understand the following:

✓ 5G networks support only the essential 5G functionalities in the first phase. The core networks are 4G-based while the operators keep adding 5G radio base stations on the field. The specifications evolve and the operators upgrade the core networks to 5G upon the availability of the new equipment, but even then, some operators may not activate all the functions.

✓ The specifications do not require 5G devices to support all the features either. As an example, the device may include only a small subset of the bands from the specified list. If the bands do not map well enough with the deployed bands of the operator, the performance may not be adequate. It is thus advisable to check that the device supports widely enough the frequency bands of the selected operator.

✓ The 5G services may be rather limited in the beginning. The operators need to install 5G radio equipment to their existing base stations or dedicate new sites, and the core networks require major upgrades. For some operators, this task may take years in order to achieve a significant population coverage.

Contract

In addition to the 5G-capable device (which may be mobile such as smart device or fixed such as FWA terminal), we need a subscription to access the 5G-network. The subscription is a contract between the customer and the operator to use the 5G network. The principle for establishing the contract does not change compared to the earlier generations; there are long-term monthly plans or lighter pre-paid subscriptions to choose from.

The monthly plan means that the operator charges the customer a fixed, periodic fee for the wireless connectivity to their network. This type of plan requires normally a commitment during certain minimum time period, which is typically 1-2 years. There may be different levels of connectivity and respective fees. A basic, lower-cost subscription may grant access to only 2G and 3G systems while a more advanced plan could provide connectivity to also 4G and 5G. Each operator decides their business model so there might be many options available.

One of the benefits of the monthly plan is that the operator may offer the subscriber with an opportunity to obtain a subsidized device with an additional monthly payment for agreed time period. As 5G devices may often be rather expensive, we may start seeing even longer-term 5G plans to lower the bar for new 5G contracts.

The pre-paid subscription refers to the access to the operator's network by paying regularly from the consumption of the services (call minutes, daily data use, number of short messages, etc.) as long as the customer wants that number to be active. If at any time the user's balance drops under a certain threshold, the telephone number is put on hold or cancelled.

Typically, the pre-paid subscription costs less for basic services but lacks of loyalty benefits such as the possibility to finance new mobile devices. The pre-paid customer may sometimes need to bring an own mobile phone to use the services.

The monthly plan customers may be able to use their own devices, too. In this case, as the device may not be certified by the operator (some operators want to dictate the minimum performance and compatibility), it is important to check that the device's supported cellular technologies and at least part of the frequencies coincide with the operator (e.g. GSM device does not work in CDMA network).

There are two types of operators; Mobile Network Operators (MNO) that own their network infrastructure and wireless operations license, and Mobile Virtual Network Operators (MVNO) that purchase part of the capacity (typically radio resources) from the licensed MNO and re-sales it to their own customers. From the customer's point of view, there are no practical differences in these two operating modes so it is matter of comparing their provided coverage areas, costs and supported services.

In 5G, all of these established models are still valid as such. If you want to sign off a 5G contract with an operator (either MNO or MVNO) that supports the new technology on the areas you want, the operator's own retail stores can provide and activate compatible mobile devices. Equally, upon the contract, you can normally activate and use your own 5G device, just making sure beforehand it is compatible with your selected operator.

Frequency Bands

The current 5G bands, and their bandwidths as per the global specifications, are listed in Figure 7 and Figure 8.

The first phase 5G networks rely on small amount of 5G base stations while the already existing 4G base stations and 4G core network support the connectivity in a parallel fashion. Thus, the 5G device can provide a somewhat faster data speeds within fragmented radio coverage areas whereas it functions still in 4G mode elsewhere.

Although there is only one specification (of 3GPP) defining the 5G networks and devices making the equipment compatible by default, the mobile phones can have many options that may not be supported by all the networks, and vice versa.

The mobile device manufacturers typically want to limit the supported frequency bands. This is a reality because each band adds up to the complexity and cost of the device. In addition, the more bands are supported, the more components such as antennas the device houses. This, in turn, increases the size and weight of the device. So, the device manufacturers tend to seek for an optimal set of supported frequencies and features for their different market areas.

To present a practical example of this principle, Table 2 summarizes the supported 5G bands of some of the 5G smart devices based on public data available on GSMArena (GSMArena, 2019). We consumers thus need to ensure that our preferred device supports some of the same frequency bands with the mobile network operator we consider to establish contract with.

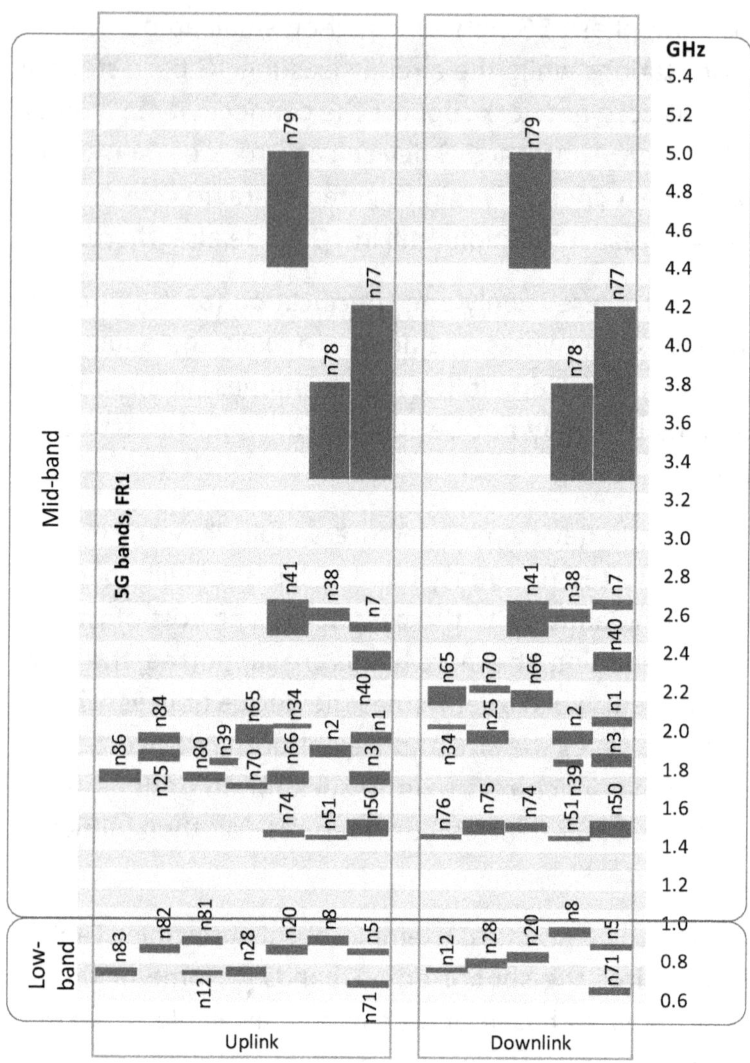

Figure 7 The low and mid-bands of 5G and their bandwidths as per 3GPP specifications. The wider the bandwidth, the more capacity and data speeds it is capable of offering. As can be seen in this Figure, the bandwidths of the low-band area are very limited while some of the allocations in mid-band offer significantly more capacity.

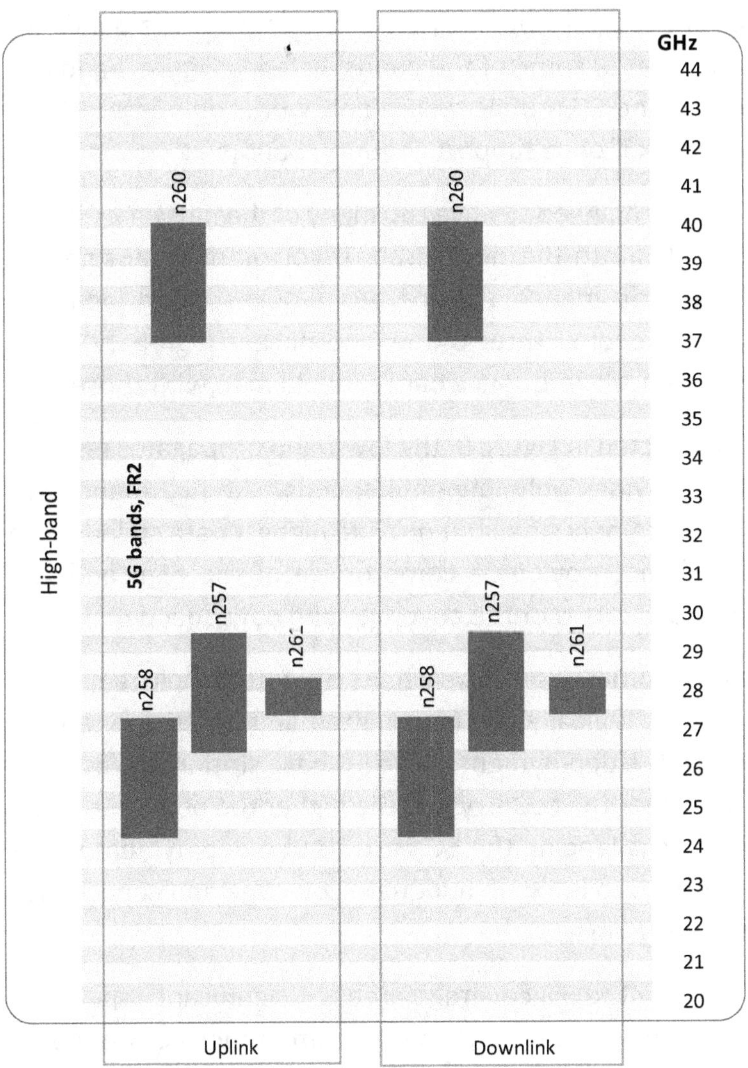

Figure 8 The high-bands of 5G. These are the new millimeter wave blocks designed for 5G. They contain much more bandwidth than the mid and low-bands do, and their ability to deliver high capacity is superior. High-bands are essential for taking full advantage of the new performance of the 5G networks. Please note that the 3GPP keeps extending the list of 5G bands in the future.

Each operator obtains license for a part of the bands (or fragment of some of the bands) listed in Figure 7 and Figure 8. The 5G mobile device only works on operator's 5G network if both support common band(s).

There is no need to have too many of the network's bands included into the device, but if you want to ensure the best performance, the operator would support ideally at least two, or all three of the low, mid and high-bands, and the device's frequency bands would map with those.

As depicted in Figure 9, the low-band propagates farthest but provides only low data speeds. The high-band cells are extremely small but provide highest data speeds. The mid-band provides a compromise of the high and low bands in terms of the coverage and data speed.

In the initial phase of each 5G operator, their radio coverage is typically still fragmented and rather limited because the deployment of the 5G base stations takes time. The larger the country and the higher the population, the more challenging the 5G base station deployment is in a fast pace to build up adequate coverage and capacity for the subscribers.

Not only the 5G coverage area is a significant aspect but the service level and data rates within those coverage areas can also vary vastly depending on the deployed capacity and the band types. It should also be noted that, within certain band, the highest data speeds can be achieved close by the base station while the speed drops gradually the farther we go from the base station.

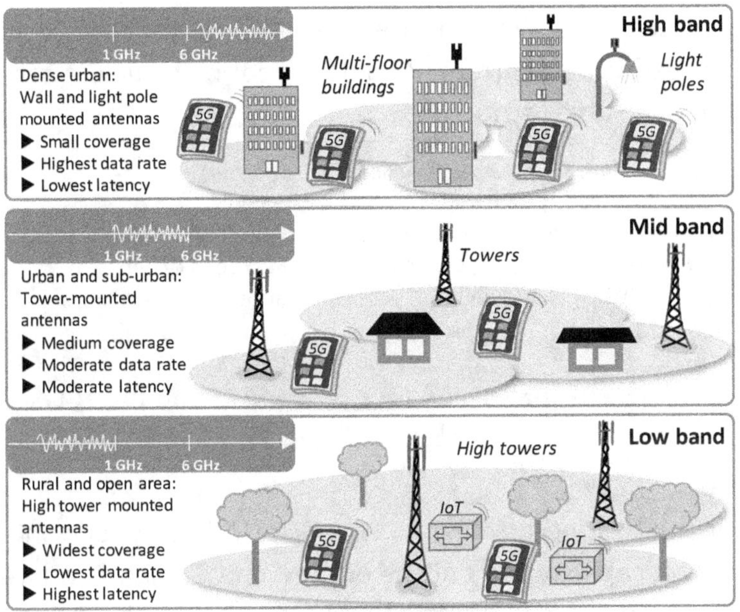

Figure 9 The key characteristics of the low, mid and high-bands.

Due to the dynamic behavior of the 5G system, it is very challenging—if not impossible—to compare the detailed and accurate performance of the devices and operators basing on only few snapshot tests on the field.

A low-band deployment provides the operator with a fast market entry and large coverage. The drawback of the low-band is its limited bandwidth, capacity and data speed, but it works well as an umbrella layer for basic services while the later deployed overlaying mid and/or high band cells can provide focused high capacity services.

Another strategy is to deploy primarily mid-band equipment. It provides a sweet spot as for the performance and coverage area. Within those areas, 5G data speeds typically exceed the current LTE data speeds making 5G's performance quite attractive.

Table 9 Comparison of the performance of the 5G bands.

Band	Frequency	Coverage	Downlink data
Low	Up to 1 GHz	Typically, ten miles or more	Few tens of Mb/s to about hundred Mb/s
Mid	1-6 GHz	Few miles	Hundreds of Mb/s to 1-2 Gb/s
High	Over 6 GHz	Within few city blocks	Few Gb/s to 10-20 Gb/s (theoretical peak rate)

For the best performing 5G, the fastest data speeds and highest capacity can only be offered on the high-band as the achieved bandwidths are superior. Due to the nature of the radio propagation, the high-band radio waves attenuate heavily as a function of distance making their cells extremely small, perhaps only city block wide. The small cells relying on high-band need to be installed densely within the desired locations such as the city centers and indoors. Due to the amount of locations and need for core network connectivity (using fiber optics or radio links), such deployment may take rather long time.

The 5G deployments have started in 2019, and thanks to the new frequency auctions, there will be more capacity and coverage available soon. The development needs to be planned well, though – not only by the operators, but the whole ecosystem needs to have the vision for ensuring optimal offering of the 5G benefits.

As stated in GSMA 5G Spectrum: GSMA Public Policy Position (GSMA, 2019), 5G needs harmonized mobile spectrum which requires "cleanup" of some of the current frequency bands by defragmenting and clearing prime bands. For the optimal performance, regulators should aim to make available 80-100 MHz of contiguous spectrum per operator in mid-bands of 5G, such as 3.5 GHz.

Furthermore, 1 GHz bands per each operator in the high-bands would ensure the high performance of the system, and the most fluent user experiences. These millimeter wave segments of the spectrum are in key position in offering the fastest data speeds and most fluent user experiences.

5G Deployment

Even if the major mobile network operators have started to deploy 5G, not necessarily all the current operators are able to advance fast to enhance the coverage. The MNOs need to prepare their network first, and it takes great efforts and investments. As an example, the major operators of the United States are all deploying actively 5G but that might not be the case for the smaller ones.

In the international environment, many operators have decided to go for commercial launches later in a peaceful way, waiting for the more developed devices and services first. As of December 2019, according to the GSMA statistics, there were 21 countries with commercially operational 5G networks. (GSMA, 2019)

During the first year of the launches in 2019, all the networks were in their initial phase which means the coverage area has still been rather limited, and the service has been based on dual connectivity mode (5G base stations connected to 4G core infrastructure), which means that the 5G performance has been still rather low.

As soon as the 5G-only Standalone mode and evolved networks will be available basing on the 3GPP Release 16 specifications, the end-to-end 5G networks start providing us with much more advanced performance.

So, in addition to the 5G mobile phones, there is a need to introduce new 5G base stations on the field. The New Radio (NR) of 5G is, as the term indicates, renewed so the previous 4G or earlier generation's base stations are not good for receiving and transmitting data with the new 5G smart devices—nor the current 4G phones can take advantage of the 5G base stations.

The MNOs can reuse their existing licensed frequencies for offering 5G, too, but most probably they want to purchase more capacity. That happens via frequency auctions that the national frequency regulators organize.

Due to all these needs for preparation, it is not straightforward to build a new 5G radio coverage fast. The time schedule depends on each MNOs 5G rollout strategy. The user's challenge is to understand the status of the coverage areas as the MNOs do not always provide very detailed maps.

The most valid comparative information for the 5G coverage and respective data speeds might thus be easiest to obtain from the technical magazines conducting snapshot field tests. Please note, though, that the results may vary greatly from one are to another, and between 5G devices.

Operator's Services

Not only the support of 5G as such, in terms of the frequency bands, is important but the services the operator offers to us consumers is equally relevant. The network itself is a kind of "bit pipe", making sure the data transmission between the user's device and the network is fast and reliable enough for supporting the features of all kind of apps, but the operator may also offer useful services that are integrated into their own network.

One of these services is the operator's voice service which in 5G is referred to as VoNR, Voice over New Radio, as per the definitions of the GSMA. The same technology may be offered commercially also via some other operator-branded name such as High Definition Voice (HD Voice).

Another useful function running on 5G is the RCS, Rich Communications Services which refers to advanced messaging.

Regardless of the availability of the specifications of these services, not all the operators support them. The basic principle is that the new 5G devices would do so by default. It is thus matter of personal preferences whether such services are important enough to consider selecting operator offering the services.

Please refer to Chapter How to Use 5G Services? for more details on VoNR and RCS.

Checklist for Selecting the Operator

Apart from the 5G connectivity and preparedness to support the desired advanced services, some of the most essential questions you might want to clarify when selecting your home operator are listed in the next sections.

Where do you need to use the device?

If you stay in a limited region such as home town, there may be local operators offering attractive cellular contracts but with limited roaming possibilities.

If you travel a lot nationally or internationally, consider a contract with an operator which has sufficiently roaming partners especially in the very areas of your interest.

As an example, some operators may provide calls and data services with the same domestic price in other countries while some others charge extra roaming costs (e.g., each chunk of data may cost a fixed price); the latter contract type may turn out to be expensive if you use (unintentionally) too much data abroad.

What is the cost of the best-fitting wireless plan?

You need to estimate the expected data consumption per charging period. The basic plans may offer a very limited total data packet per month while the extra data costs more depending on the use. Another option is so called unlimited data plan which provides maximum data speed, but typically only up to certain limit. After this limit has been reached within the billing period, the data speed is reduced until the next billing period starts.

For the heavy data users, especially with the higher data speeds of the 5G systems, the currently utilized basic plan can run out of the default data rather soon which may result in more expensive monthly payment compared to the unlimited 5G data.

There may be bundles for the wireless plans combining other services such as Wi-Fi and TV resulting in more attractive price compared to each individual contract, if you use all those services. Please note that sometimes the hidden costs may bring surprises, such as setting up the service, optional insurance (which sometimes is added into the plan by default even if you do not need it), etc., so it is worth clarifying the costs as detailed as possible beforehand. It also should be noted that the price typically increases after the first contract period ends.

Quality of the service

Some operators may have occasionally lack of customer service resources. The level of the response times and the quality of the customer service can be investigated from different web pages reading through the crowd-sourced opinions. As is the case in any other crowd-sourced information, and as a rule of thumb, it is a good idea to discard the most critical and overwhelmingly positive opinions out first, and understand the voice of the majority.

The quality of the network itself is one of the most significant selection criteria. The quality refers to the offered data speeds especially within the area of your own interest, to the coverage, and to the continuum of the service.

The ways to investigate this is limited as the operators typically would not reveal very detailed coverage and respective data speed maps. It is hardly enough to observe a high-level map indicating the presence of the 5G while the quality (meaning coverage, data speed and capacity) can vary so greatly. As an example, if the operator has only limited capacity in certain area, the single connections suffer as the number of the simultaneously communicating customers increase in that area.

One way to understand the quality is to investigate yourself by performing data speed tests, or rely on overall tests that can be obtained from public sources. As an example, the GSMArena summarizes performance values of various 5G handset models. It is a good idea to compare the evaluations by seeking for alternative sources, too, such as the customer feedback of different online retail shops.

How to Set Up the 5G Phone?

5G mobile device refers typically to a smart device for us users. It can have many other forms, too, from cellular-connected small sensor or sports tracker up to the most complicated multimedia and multi-use device. In this context, we assume you are considering to obtain, or have already purchased a 5G smart device.

Operating System

5G as such does not change too much the overall looks of the typical smart device we already have got used to. Its form may vary from the most basic phone up to high-end, highly sophisticated device capable of performing many other tasks apart from the pure 5G connectivity.

In all of the cases, the phone needs an operating system in order to process the calls, data transmission, and run the applications. The typical operating systems are going to be the very same as has been the case in previous generation's LTE phones, the two market leaders being Google's Android and Apple's iOS.

So, you might first want to decide whether you want an Android device or iPhone. If you are happy with your already established services and respective OS of previous, LTE or other cellular device, the benefits of going for the same in 5G include the already familiar user experiences, and the easiness to synchronize your previous data such as contact list between the devices supporting the same OS. The synchronization typically takes place via cloud.

Activation of the Device Using SIM

When we get a brand new 5G smart device, it needs to be activated. 5G devices still rely on the already established SIM card (Subscriber Identity Module) which may be removable as has been the case in the previous generations. There will also be increasingly popular base for devices which integrate the SIM functionality into the device's hardware and relies on the eSIM concept.

If the 5G device supports the traditional, removable SIM card, it is inserted into the SIM tray when the phone is powered off. Once the SIM card is in, the mobile phone can be switched on. The process from this point forward is exactly the same as has been with the previous generations. The operator that provided you with the SIM card needs to provision the subscription when the SIM is used first time. The initial provisioning happened already when you obtained the SIM card, as the operator made a contract with you to use the mobile network and added your information to the 5G register. The 5G network thus knows to wait for the SIM card to report its intention to start using the network resources. Once the 5G device is powered on, the signaling triggers the final provisioning procedure between the device and network's home register. At this point, you need to type in the information as requested on the screen, and as soon as finalized, you are good to start using the device.

The more modern way is to activate the device without a removable SIM card, assuming the 5G device contains an embedded variant of the SIM in the device's electronics. The activation can take place in many different ways from which one of the currently used is a QR code (the two-dimensional barcode familiar from flight tickets). By scanning the QR code with the new 5G mobile phone, it triggers the provisioning and subscription's activation procedure, requesting the needed customer's input on display.

If you purchase the mobile phone in your own MNOs retail store, the customer service will take care of the activation of the 5G subscription. It can be included as an additional service to the existing subscription, maintaining the old mobile number, or there can be entirely new subscription for the 5G.

If you purchase 5G device in open markets, the customer service of your home operator needs to include the 5G subscription into your customer profile either visiting the retail store, or by calling to the customer service.

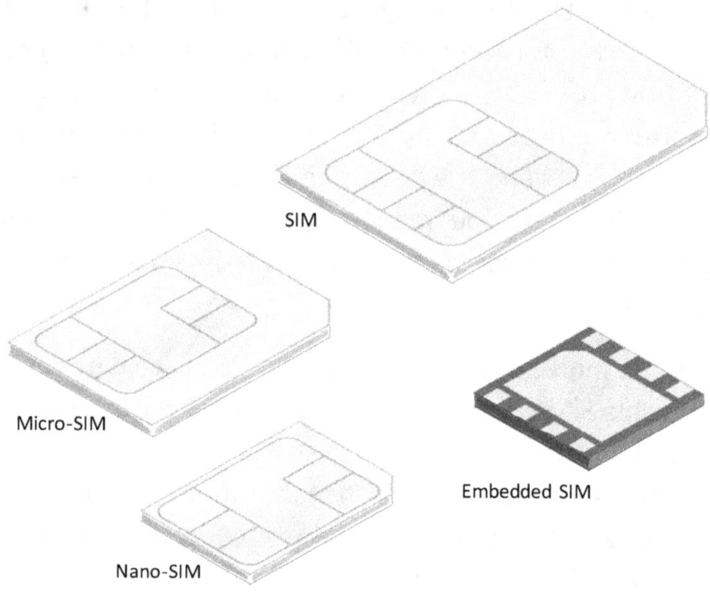

Figure 10 Removable SIM cards and embedded, non-removable SIM.

It should be noted that even if the mobile phone relies on the traditional SIM card, the 5G subscription may require a physical upgrade of the (more performant) card especially when the MNO starts to support the fully end-to-end 5G radio and core networks. Secondly, the renewal of the SIM card is a good idea in any case because its memory tends to wear out gradually (each SIM has certain maximum number of read and write cycles it tolerates during its lifetime). You can request your operator to deliver the new SIM via postal service or you can swap it at the customer service point of your home operator.

eSIM Activation

eSIM is a new addition to the SIM family. It is a "virtual" version of SIM that makes it possible to activate a cellular plan from a carrier without having to use a physical SIM card. The eSIM can be based on an embedded SIM hardware that is soldered permanently into the device, as shown in Figure 10, but it also can work on the traditional, removable SIM.

If the device supports eSIM, the plan activation procedure differs from the one that has been used with the physical SIM card earlier. In this case, you do not need to insert new SIM card into the device. The activation takes place by selecting the appropriate options from the device's menu. Upon the request, the plan is activated automatically over the air (OTA).

In any of these above-mentioned cases, the subscription is activated and the 5G device can be used as of now for high data speed connections wherever the 5G radio coverage is available. If the 5G radio network has outage, the device and network automatically hand over the connection to operator's other radio network that is found in the coverage area.

There will be increasingly versatile 5G devices in the markets including smart devices, companion smart watches and fitness devices, many different sensors and other machine-to-machine devices, and they rely increasingly on the embedded SIM. In order to work in 5G mode, they all need the activation of the subscription.

One of the practical use cases is related to the companion devices such as sports tracker. The idea is to use the respective primary device such as smart device in a daily basis. When you go jogging and want to leave the smart device home, you might want to be able to still use the basic functions such as voice calls and text messages on the go which may be possible with the sports tracker supporting those services via its own cellular connectivity.

In such a case, the sports tracker may have an embedded SIM and the same subscription can be switched from the smart device temporarily to the sports tracker for the jogging period. The procedure depends on the operator, e.g., by activating the subscription on the smart device or the companion device via the menu of the provisioning service app. The subscription can be thus swapped between the devices on-the-go via the menu.

The eSIM is a global specification by the GSMA which enables remote SIM provisioning of any mobile device. The eSIM allows consumers to store multiple operator profiles on a device, and switch between them remotely. (GSMA, 2019)

The eSIM support will be increasingly popular, and can be assumed to be important in the future 5G devices to complement, and potentially at some point even replace the SIM cards. There may also be more devices coming into the markets supporting solely eSIM without even option to use removable SIMs although this evolution depends on the overall adaptation of the concept.

The activation is typically done by utilizing an operator-specific QR-code which is scanned and which instructs the respective operator to trigger the activation for that specific mobile device.

For the devices supporting the eSIM, the following examples show how some operators handle the eSIM activation when using eSIM-capable device. The user can swap the operators simply selecting the desired one from the menu of the device.

Please note that the device can be the primarily utilized smart phone as such or a connected, companion device such as smart sport tracker. The same activation principle applies globally for majority of the other operators, too, if they only support the eSIM concept from the network side, so please refer to the more specific support and instructions of each operator at their web pages or by contacting their customer service.

▶Verizon

Verizon's eSIM activation instructions can be found at their web page.[22] At the moment, the eSIM activation can be done with a set of iPhone models.

Prior to the activation, make sure you have a data connection which can be Wi-Fi or cellular plan from another wireless carrier. Also, you'll need to get Verizon service for your eSIM.

---

[22] *https://www.verizonwireless.com/support/esim-activation-instructions/*

✓ New customers may sign-up for Verizon service which happens downloading the "My Verizon app" (from the App Store of Apple), and following its instructions to confirm eligibility and set up the account.

✓ For the ones with previous Verizon service using your iPhone's physical SIM card, you'll need to contact Verizon's customer services to add service to the eSIM.

Once you have signed up for Verizon service, you'll need to setup the iPhone supporting eSIM. This happens by displaying the QR code below (in any optical means including this page, computer or tablet), and scanning it on the iPhone by tapping "Settings" – "Cellular" – "Add Cellular Plan". Please use the camera or an app on your smart phone that scans QR codes.

eSIM activation of Verizon

Figure 11 The eSIM activation code of Verizon.

In case your phone requests to select a default cellular line, the new Verizon line can be selected for voice, messaging and data.

After this, the installation completes, and the "Cellular Plan Added" notification is displayed and you are ready to start using the new Verizon service.

▶T-Mobile

To activate an eSIM on T-Mobile network, you need to visit a T-Mobile retail store to activate a postpaid line on an eSIM.[23] This ensures the device is eligible for the setup.

✓ For the T-Mobile postpaid customers, using an eSIM-capable device which is based on a removable SIM card and desiring to swap the service to an eSIM, T-Mobile advises you to visit a T-Mobile store or contact the customer care for assistance.

✓ In case you want to activate a new prepaid account on iPhone with eSIM, you can download and install the T-Mobile "Prepaid eSIM" app.

In order to download the eSIM settings, the eSIM activation can thus take place either at a T-Mobile store or by swapping the existing SIM for an eSIM.

To proceed doing so, you'll need an access to the Internet on your device either via Wi-Fi or other means. In case the physical SIM is changed to an eSIM, please remove it from your device. Now, you can download the eSIM settings using the QR code below. As an alternative way, you can also type directly the path to the T-Mobile eSIM service[24], which is "SM-DP+ Address": T-MOBILE.GDSB.NET

---

[23] *https://support.t-mobile.com/docs/DOC-40796*

[24] The term SM-DP+ refers to "Subscription Manager, Data Preparer" which is defined by the GSMA as a part of the operator's remote SIM provisioning platform. More technical information on the eSIM can be found at: *https://www.gsma.com/esim/* and *https://www.gsma.com/esim/esim-specification/*

*eSIM activation
of T-Mobile*

Figure 12 The eSIM activation code of T-Mobile.

Please note that the Prepaid eSIM app asks also the "Activation code" and "Confirmation code" which are left blank. After these steps, please find the menu items to choose your device and select eSIM settings in the Connection & Network drawer.

✓ For iPhones with eSIM, go to Settings > Cellular > Add Cellular Plan.

✓ For Pixel 4 & Pixel 4 XL, go to Settings > Network & Internet > Mobile Network.

Now, the eSIM activation is ready.

▶AT&T

AT&T instructs the users about the eSIM activation at their web page.[25] For iPhone XS and newer models, as soon as you have the cellular service plan set, you need to download the eSIM. In case you want to bring your own device or add a second line, you need to make order for another online. If you want a new service, a new phone, or an upgrade, contact the company or visit a retail store.

---

[25] *https://www.att.com/support/article/wireless/KM1008711*

You might find your eSIM activation QR code printed on an AT&T eSIM card. In that case, find the phone's settings. Depending on the device, select one of the following: 1) "AT&T Cellular Plan Ready to be Installed" – "Continue"; or 2) "Cellular" – "Add a Plan".

In order to test that the activation was successful, check that the SIM card is inserted correctly and the device is on, and place any call or browse the web (deactivating Wi-Fi during this test). If you experience problems, please activate your device online at *www.att.com/activations*.

▶Other Examples

Table 10 collects information sources on some the operators supporting the eSIM activation.

Table 10 Some operators providing eSIM activation and the links for the respective instructions.

Operator	Web
Sprint, USA	*https://community.sprint.com/t5/Moto-Google-Essential-Others/How-to-activate-eSIM-on-Pixel-3/td-p/1040476*
Truphone, USA	*https://www.truphone.com/us/iphone-dual-sim-plans/*
Ubigi, USA	*https://cellulardata.ubigi.com/esim/*
Bell, Canada	*https://support.bell.ca/mobility/smartphones_and_mobile_internet/what_is_dual_sim_and_how_do_i_use_it?step=2*
Telefonica, Spain	*https://www.movistar.es/particulares/movil/esim*
Telia, Finland	*https://www.telia.fi/kauppa/liittymat/e-sim*

The more complete list of the current operators supporting the eSIM can be found at the web page of Apple.[26]

The Google ecosystem is getting traction, too. As for the models beyond smart phones devices, there have been already some commercial companion devices supporting eSIM such as Samsung Gear S2 and S3. These additional models can be used for stand-alone phone calls, too, once provisioned.

It can be expected that there will be more eSIM-capable Android devices introduced during 2020.[27] Also Google has already updated their ecosystem to include eSIM-capable smart devises such as Google Pixel.[28]

At present, Apple has iPhone models that support dual SIM with an eSIM. According to the web page of Apple[29], iPhone XS, iPhone XS Max, and iPhone XR equipped with the iOS 13 support this feature.

The dual SIM can be utilized for using one telephone number for business and another for personal calls, or to add a separate data plan when traveling outside of the country or region. The eSIM concept also allows the user to have separate voice and data plans.

---

[26] *https://support.apple.com/en-us/HT209096*
[27] *https://blog.gemalto.com/mobile/2019/02/20/how-android-the-worlds-most-popular-mobile-os-is-preparing-for-esim/*
[28] *http://justswipe.com/googles-esim-manager-app-now-play-store/*
[29] *https://support.apple.com/en-us/HT209044*

Setting up the 5G Mobile Device

As soon as the 5G subscription has been provisioned as explained above, the device can be activated. This is typically straightforward task as the settings are most likely correct by default.

Prior to the use, it is a good practice to make sure the device's battery is fully charged. Once ready, turn on the power.

The first time you do this, the setup wizard of the device is displayed and walks you through the key settings. This phase makes it easy to select the default language and Wi-Fi network, and to set up the accounts; in case you do not have yet accounts, the vizard instructs how to obtain one. You can change and update these selections later from the settings menu. Each model has its own menu structure so it is a good idea to read it through, also to learn the special functions and supported, oftentimes overwhelmingly complete set of services.

At this time, the only 5G specific setting is related to the 5G access; the device will attempt to attach to the nearest 5G cell by default, and once the network recognizes the new subscription and device, the 5G logo will appear to the display. If this does not happen, the most probable cause is the lack of 5G coverage on that area, in which case the device selects some other available system of your home operator. You might want to double-check the cellular network settings on your device, though, to make sure the cellular data is enabled.

Setting up the Fixed Wireless Access

The 5G Fixed Wireless Access (FWA) refers to a customer premises equipment (CPE) that is installed permanently within the preferred location such as home or office. It can be compared to a Wi-Fi router—and in fact, it is the same, the only difference for us consumers being the connectivity of the router further to the infrastructure by 5G radio instead of the fiber optics or coaxial cable.

The 5G CPE houses a 5G radio equipment and can have connectivity with a Wi-Fi router. The CPE is installed typically in some indoor location where the 5G radio signals are best received, such as close by or on the window.

The FWA access equipment communicates with the operator's nearby 5G base station, and can relay the traffic to a wireless Wi-Fi 6 router (as per the IEEE 802.11ax standard), providing a high-speed data connection for the customer connected to it. Commercial examples can be found, e.g., from Verizon's Home Internet web[30].

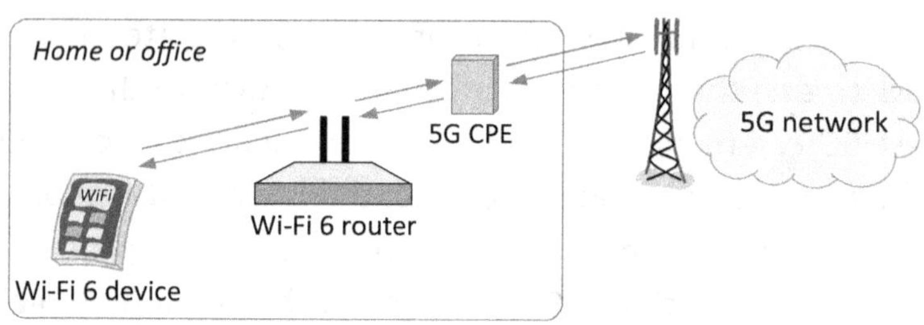

Figure 13 The principle of the FWA.

---

[30] *https://www.verizonwireless.com/5g/learn-5g-home-internet/*

How to Use 5G Services?

The 5G smart device is capable of performing all the tasks that the devices of the previous generations have been able to do, including voice calls, data services and text messaging. Nevertheless, the methods are renewed technically.

Data

The obvious benefit of 5G is the increased data speed. Not only that, but the advanced version of 5G (referring to the 3GPP Release 16 -capable devise and networks) have also other benefits such as very low latency and better reliability, where applicable.

The utilization of any of these basic services does not differ from the previous generations. The data transmission is transparent to the user, and selecting, e.g., YouTube video, the only difference the user sees is the faster downloading of the contents meaning there will be less breaks in the video. The 5G thus fills the buffer of the video player much faster than the previous networks do.

This same principle applies to any other application which is based on fast data transmission and reception. 5G may provide so fast data connectivity that the potential bottleneck may occur elsewhere, not on 5G radio interface. Example of such occurrence is a public Internet and its server which may not support the maximum 5G data speeds, so the user may still see lower data rates regardless of the available 5G radio capacity depending of the type of service and its supported data speeds.

It should be noted that especially during the initial phase of 5G, the operators will deploy increasingly more capacity on the radio interface while the actual services might not necessarily always follow the development. Thus, even if the smart device and the cellular network may offer significantly better data speeds, the servers at the other side might not quite yet be capable of offering such level. This is one of the realities in the beginning, but as time passes, we can expect to get more consumer services which are compatible with the 5G capabilities.

It also should be remembered that 5G is not only about higher data speeds but the other benefits include the support of vast amount of simultaneously communication devices which benefits especially the low bitrate machine-to-machine type communications, and ultra-reliable connectivity which is good for critical communications such as autonomous vehicles.

Voice and Video Calls

Voice call is still one of the most important applications in 5G. As 5G does not support the old-fashioned circuit switched voice calls, the voice calls are based on packet data as defined by the Internet Protocols (IP). This Voice over IP in 5G is called VoNR, Voice over New Radio.

VoNR is a similar than the VoLTE of the 4G networks (Voice over LTE), but it is further tailored for 5G. The technology remains transparent for consumers, and the VoNR is present via a voice call application as soon as available.

Both the 4G VoLTE and 5G VoNR means that you can make high-quality, high definition voice and video calls within the cellular network. They are integrated services so there is no need to download a separate app as is the case for the separate, Over the Top (OTT) applications such as WhatsApp, Messenger and Skype.

If only the home operator supports VoLTE or VoNR and has activated the service, the service may be already set. The operator may not necessarily use the term "VoNR", though, but some other name such as High Definition Voice and Video. Depending on the model, the activated and available VoLTE or VoNR can be seen as an icon on the top of the display, such as in Figure 14.

The user can activate and disable the use of the VoLTE and VoNR from the settings of the 4G and 5G phones. The setting depends on each mobile phone model and operator. As an example of the variants, an Android phone might have one of the following paths to toggle the feature on and off:

Settings > Connections > Mobile Networks > VoNR calls

Settings > Connections > Advanced calling

Figure 14 The available VoLTE and VoNR are indicated by their (device-dependent) symbols on the mobile phone's display.

The main difference of the VoLTE / VoNR and the Over the Top voice applications is that VoLTE and VoNR are integrated services within the 4G and 5G infrastructure so there would be no need for the customers to download a separate application.

When calling to other subscriber, the number is typed on the telephony or messaging app. The numbers can be stored on the memory of the device, or they can be assigned on some of related accounts such as Google. The benefit of the latter is that the information can be saved and synchronized between any mobile phone.

When you store the mobile telephone numbers in the phone book of the device, it is recommendable to use the international format (starting with "+" and country code, e.g. "+1" in the USA). In this way, you do not need to retype the numbers in the international format when calling from abroad.

The 5G networks and devices serve especially well the video applications as they require typically high data speeds, yet the latency or its variation (jitter) is of no issue in normal video streaming. The streaming happens utilizing a mobile app that connects the services such as YouTube or two-way video conferencing.

Messaging

Instead of the original Short Message Service (SMS), the default messaging platform in 5G is the RCS, Rich Communication Services. Like VoLTE, the RCS has been already deployed in many LTE networks as it is designed to replace the somewhat more limited SMS.

The RCS brings along the interactive messaging, and in addition to the peer-to-peer, direct messaging between users, there are business messaging and chatbot-messaging options, too. The enrichment includes also embedded graphics, location, user status, typing indicator, and many other functionalities that are similar to the OTT applications. Again, the benefit of the RCS is that it is in-built within the 5G infrastructure so the consumers do not need to do anything to install or activate it.

The challenge of the RCS, as is the case with the VoLTE, is that the network and mobile phone alike need to support it. At present, on January 2020, the statistics indicate that there are 84 RCS operators worldwide. The GSMA forecasts an additional 24 operator launches by the beginning of 2020. (GSMA, 2019) Typically, the new mobile devices support RCS, so it can be assumed that the 5G smart devices do have the RCS support by default.

Special Services

5G introduces also a variety of special services. Some examples include train communications (replacing GSM-R), vehicle-to-vehicle (V2V), and satellite component.

One of the interesting 5G service is the direct communications, or Device to Device (D2D) mode.[31] It may turn out to be an important base for many applications in consumer markets and special communications alike. Its idea is to facilitate 5G mobile devices to communicate directly relying on a control link between 5G base station and a device. Some potential D2D applications are local content sharing, public safety, underground connectivity and vehicular communications.

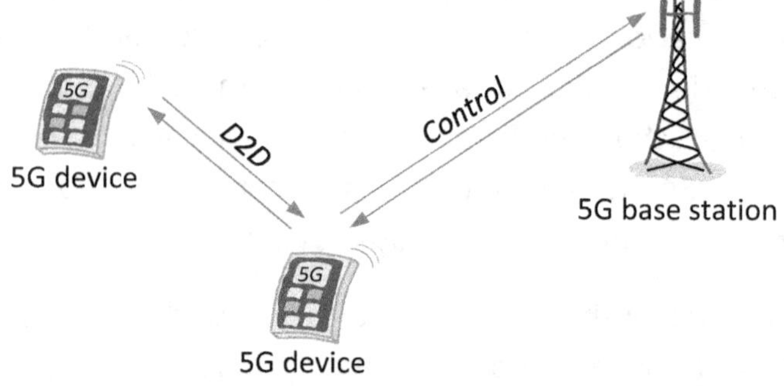

Figure 15 The principle of the Device to Device (D2D) communications.

---

[31] **http://www.cs.tut.fi/kurssit/ELT-53207/lecture14.pdf**

What Added Value 5G Offers?

One might argue that the networks do not support yet 5G-specific applications. Developers might not have yet new, interesting solutions either while the 5G networks and user devices are in their infancy. Furthermore, even if the 5G networks and devices, where available, may provide increased bit rates, the rest of the telecom infrastructure, including public Internet and servers behind it, may not yet be always able to support such speeds.

This is a typical chicken-egg challenge. Like in any other environment, there will be forerunners solving the issues. That creates plenty of new technical and business opportunities for many existing and completely new stakeholders. Some examples of the opportunities include the housing of 5G network functions in data centers provided by external entities. Another opportunity is mobile operator's own data center operating 5G core, which can expose some functions such as location-based services, or secure cloud storage with mobile authentication, also to third parties extending the current business models and customer base of the MNOs. It is thus matter of time to see these new businesses arising as soon as the new ecosystem will be up and running.

Meanwhile, and gradually, the basic task of MNOs is to deploy the 5G networks and form the new infrastructure.

Development of the Ecosystem

Due to the vast business opportunities relying on the new, more performant platform for the delivery of more capacity faster and with lower lag, it is obvious that the 5G industry is willing to go forward as soon as technically feasible. The elemental step in this process is the setting up the new, standard-based networks and equipment in the global level.

All the essential stakeholders have demonstrated the interest on 5G, including chipset and module makers, network and device manufacturers, service providers and mobile network operators. There are also plenty of existing and new verticals expressing their needs for the future 5G networks to help executing more efficiently their functions. As envisioned by the ITU in their connected society concept, 5G will be an elemental and important communications and data transfer platform for a whole society including smart cities, critical infrastructure, telematics, and a huge amount of IoT devices, to mention some. That, in turn, will enhance the ecosystem with its infrastructure, services, and applications.

As the 3GPP specifications evolve and the respective devices and solutions will become available, there will be totally new applications developed, too. This is the familiar pattern we have already seen throughout all the previous generations since the very first one back in 1980s.

For us consumers, the clearest benefit of 5G in the beginning of the 5G era is the increased data speed. We can download contents faster, starting with close to one Gb/s in the beginning of the 5G services whereas the theoretical single-user download speeds will be up to 20 Gb/s as dictated by the ITU IMT-2020 for the respective evolved broadband use cases.

The initial networks, aka 3GPP Release 15 -compliant installations, are still largely based on 4G infrastructure (core network). This phase uses the intermediate Non-Standalone options combining 4G and 5G communication links.

The further evolution of the 5G networks will bring along the Standalone architecture, which is the ultimate goal of the operators. It includes only 5G radio and core networks, and it will be able to offer finally the fullest performance of 5G to us consumers.

The low latency is definitely of interest to the ones requiring real-time user experiences. Such fast response times are highly beneficial in the Augmented Reality (AR) and Virtual Reality (VR) games and other applications. Augmented Reality means that the device superimposes a computer-generated virtual image on a user's view of the real environment This composite view is an interesting technology area which can be used, not only for gaming but for many type of environments mixing the computer-generated and real views such as in car while the accompanying information is superimposed on the windshield.

Combining the 5G system's low latency with the high data speed and high reliability, such applications may become popular, including vehicle communications that also provides advanced entertainment and control systems.

Beyond the vastly increasing data speeds and the low latency for special environments, there are many verticals—user types—that may benefit from 5G even more than we ordinary consumers. The combination of the eMBB, URLLC and mMTC will be highly relevant in such environments as Smart Cities and the ones using advanced cloud edge services.

Enhanced Functionality

5G changes completely the traditional network architectural model. Whereas the traditional stand-alone network elements rely on their own software per device, the 5G network is based on a common hardware and selected software instances running on top of it and serving different 5G network functions as per need. This optimizes greatly the performance of the network and makes the adoption of new services, such as edge computing, possible much faster than in the "traditional", non-virtualized networks.

5G core and radio network functions can be processed in the cloud environment of the data centers, so we will see an increasing number of cloud providers offering their capacity to the mobile network operators.

As soon as these advanced functionalities will be available for us consumers, it is possible to take advantage of the low latency and high data speeds in much more efficiently than has been the case with the 4G networks.

How Consumers Benefit From 5G?

As 5G networks and their evolved services will become more widely available, there is going to be an increasing number of us consumers as well as completely new user segments, or verticals, benefiting from the high capacity and data speed, low latency and ultra-reliable connectivity of the 5G networks. Such verticals include many users such as law enforcement, transport, and health care, as well as vehicle communications, and augmented and virtual reality applications.

Further examples of the other segments include the communication and control of drones, remote and secured communication of medical equipment, factory automation, and enormous amount of IoT sensors.

The basic question for these segments is what added value 5G will bring over LTE. We users might not need yet performance beyond the current LTE, say, for using social media and consuming multimedia such as YouTube.

Of course, many of us would be happy to be able to download any data in seconds instead of minutes via eMBB, or use lag-free AR/VR via URLLC, so it is also matter of rising the interest of the developer community to create new apps and services that take advantage of the lower latency and faster data—together with the gradually evolving 5G coverage and capabilities. It may be assumed that the related industry initiatives such as the 5G Operator Platform can expedite this evolution and solve the chicken-egg challenge.[32]

Table 11 summarizes some of the foreseen use cases of the 5G networks for the consumers.

Table 11 Some potential uses of 5G for consumers.

Use case	Benefits for consumers
File download	Much faster data transfer from servers to the 5G devices or companion devices such as laptops. 4K movies can be estimated to be typically around 100 Gigabytes in size. The downloading of such contents via 5G, in the best case of 20 Gb/s speed, would take around (100 GB x 8 bits) / (20 Gb/s) = (800 Gb / 20 Gb/s) = 40 seconds. In the beginning of 5G, the practical download speed may be around 2 Gb/s which means that such contents can be downloaded in 400 seconds, or less than 7 minutes. In comparison, 1080p High Definition (HD) videos are typically approximately 4-8 GB in size which takes only few seconds to download via the theoretical 20 Gb/s speed of 5G, or half-a-minute in the realistic first-phase 5G networks.

---

[32] **https://www.gsma.com/futurenetworks/5g-labs-seminars/**

Streaming	Services such as YouTube transfers only part of the data and keeps buffering more while the consumer is streaming the contents. 5G will further minimize the breakdowns due to the failing in downloading. 5G can also provide more optimized cloud edge for storing the desired contents closer to the user.
Smart home	Apart from the home Wi-Fi routers, the appliances can also benefit from 5G connectivity. One example of the evolved smart home could be the use of 5G Fixed Wireless Access (FWA) for 5G-connected TV, intelligent sensors and AI-equipped robots. The integrated security of the 5G networks might be highly beneficial to offer an extra layer of protection against malicious actors.
Automotive	5G is designed to serve also as a platform for vehicle-to-vehicle communications (V2V) and autonomously driving vehicles which translates to the better quality of life for us consumers.
Games	5G can tackle the demanding requirements of the advanced gaming thanks to the combination of the 5G pillars, eMBB and URLLC.
Multimedia	5G is a logical base for mobile multimedia which can be applied to countless environments, including entertainment on-the-go. At the same time, the traditional consumer devices may take many other forms such as foldable screen, and are expected to include even holographic projection and 360-degree cameras in the long-term future.
AR/VR	Augmented Reality and Virtual Reality applications have been defined as some of the primary use cases of 5G. Not only the gaming industry is expected to offer novelty applications and products basing on 5G connectivity and AR/VR, but there may be plenty of other services such as virtual shopping via mobile (comparing different clothes and other products cozily from home).

More information about consumer use cases can be found, e.g., from the Ericsson Consumer & IndustryLab Insight Report. (Ericsson, 2019)

Use Cases for Verticals

A use case refers to a certain situation in which a product or service could potentially be used. The evolved Mobile Broadband (eMBB) serves as a base for use cases that require fast data transfer, such as large file downloading, audio-video streaming, and multimedia messaging. These use cases are not necessarily critical for the delay while the information is sent or received in timely manner. The eMBB use cases are the ones that we users and verticals alike can start experimenting during the first phase of 5G.

The second phase, referring to the Release 16 -compliant 5G networks, will provide many new non-eMBB use cases as the capacity per site increases and the new massive Machine Type Communications and Ultra Reliable Low Latency Communications are available. These use cases can also rely on the cooperation between 5G and wired environments such as Time Sensitive Network which is highly reliable counterpart for the mobile 5G.

An example of such use cases is telemedicine that could provide end-to-end connectivity on rather wide areas for the most critical remote surgery. Another example is a remote manufacturing control in a safe and reliable manner within and between critical production sites.

Today, these use cases are still merely ideas while operators upgrade their 5G infrastructure and interconnect the 5G radio and core to the new wired infrastructure. Once this work is completed and there are concrete applications and services enabled and available, the verticals can start taking advantage of the mMTC and URLLC modes.

At the same time, these new solutions can serve as a platform for completely new businesses that we might not be able to even visualize quite yet until the enablers are in place.

5G is able to support a variety of use cases independently from each other. Each of these use cases require different performance values in terms of area traffic capacity, peak data speed, user's experienced data speed, spectrum efficiency, mobility, latency, connection density, and energy efficiency.

The offering of these capabilities with an adequate quality of service (QoS) per each vertical, or user type, is possible thanks to the new 5G technologies such as virtualized 5G network functions, software defined networking, edge computing, and network slicing. The latter means that a 5G network can be segmented so that within the very same radio coverage, different verticals may select the most adequate from a variety of 5G network types.

There are many existing and new verticals that benefit from such personalized service. The following sections summarize the opportunities for some of them.

▶SME Business

5G may offer many new benefits for the small and medium enterprises. Some of the most logical use cases are related to the evolved payment processes that can rely on 5G mobility in a secure manner, fortified by additional layers of security. Furthermore, the 5G system is capable of working as a platform for virtual tactile shopping, drone delivery, hologram calling, and 5G facial recognition payment, to mention some examples of the analysis by Ericsson. (Ericsson, 2019) Within the SME business, the 5G network can serve as an evolved version of the company's internal PBX. This can be done by dedicating a network slice for such functions, or by deploying a network segment relying on unlicensed bands.

▶Transport and traffic

5G can help optimize traffic flows by offering fast and reliable communications and control links between vast number of sensors, surveillance cameras and other devices connected to Smart City traffic management systems.

Combined with Artificial Intelligence and Machine Learning, adaptive prediction models, and analytics based on environmental variables such as weather forecasts, predicted mass events that impact transportation flows, and seasonal traffic histograms, the 5G can provide a highly feasible and beneficial component to this ecosystem.

There are many research programs set up on the topic. As an example, the smart highway in Atlanta, Georgia, works as an outdoor laboratory to tests future solutions, including the V2X communications. These intelligent road solutions are still rather far away in the future but 5G can well form part of such concepts. (Quain, 2018)

▶Vehicle communications

The 5G system can serve as a service for Vehicle-to-Vehicle communications (V2V) between cars, Vehicle-to-Infrastructure communications (V2I) between the vehicles and networks, and Vehicle-to-Everything communications (V2X) within the while ecosystem.

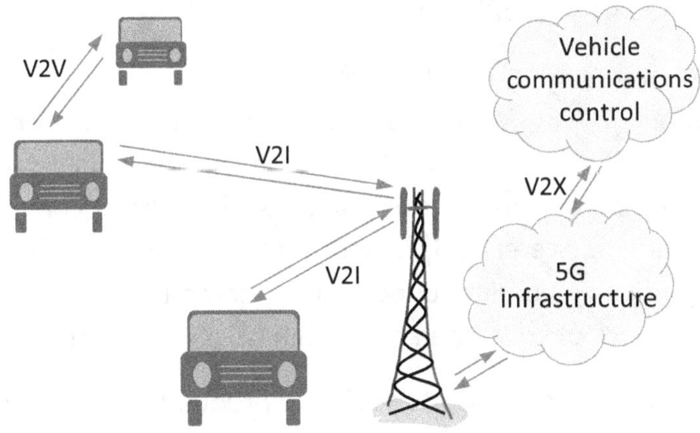

Figure 16 5G serves as a platform for vehicle communications.

Combining the URLLC-based communications of 5G with vehicle's internal Augmented Reality functions, the new automotive ecosystem may turn out to be highly interesting, and may include 5G-based AR-projected windshields, autonomously driving cars, and advanced in-car entertainment systems. 5G could also offer a platform for further developed concepts such as see-through cars. As an example, 5G Automotive Association (5GAA) has investigated the benefits of 5G and considers it to serve as a feasible platform for advanced vehicle communications. (5GAA, 2020)

The New Radio of 5G can enable V2X communications via the 5G infrastructure, including MBMS (Multimedia Broadcast Multicast Service) support. It should be noted that the 5G's V2X is not planned to replace the V2X services offered by LTE, but they have designed to complement each other, and they are capable of offering interconnectivity. (3GPP, 2019)

▶Manufacturing and Industry

There are many industry areas benefiting from 5G in their daily tasks such as remote operation and control. 5G tackles also some very special aspects such as direct communication between devices which is beneficial in, say, mining industry needing underground communications due to the lack of coverage area within such special locations.

Relying on a local 5G network, the production site can reduce cabling costs and still ensure a highly reliable remote connectivity. 5G also can support autonomously moving units such as factory carts in logistics center.

There is an industry trend to merge the Time Sensitive Networks (TSN) and wireless world. TSN refers to the evolution of wired Ethernet based on IEEE 802.3 and IEC/IEEE 60802 standards making them more reliable and faster while reducing latency. TSN can be applied to industrial automation, vehicle networks, and many other environments that require enhanced performance.

Thanks to its low latency and ultra-reliable characteristics, 5G is a logical component to be integrated into this concept. It can connect wireless devices such as intelligent industrial sensors and reduce the need for cable installations. This concept will be feasible along with the new modes of the 3GPP Release 16. (Ericsson, 2018)

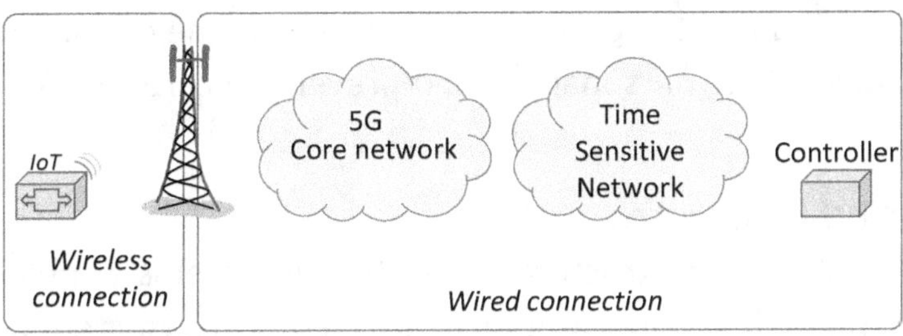

Figure 17 The merging of 5G and Time Sensitive Networks can benefit industrial verticals.

▶Health Care

5G can provide an important communications link for the people in rural areas to save invaluable time in setting up and visiting medical appointments. The role of the 5G network would be to provide means to perform telehealth and remote home monitoring, and is beneficial for doctors in their efforts to assess the patient's state via high capacity video call. In such environment, the latency is not critical but the quality of the connection, including possible imaging contents is of utmost importance in order to ensure correct diagnostics.

AT&T has identified important aspects of 5G which can work as a platform in health care environment. The benefits of the 5G system include the ability to quickly transmit large imaging files, provide telemedicine, enable enhanced AR, VR and spatial computing. It also facilitates reliable, real-time remote monitoring and the integration of Artificial Intelligence. AT&T has concluded that by enabling the advanced technologies through 5G networks, healthcare systems are able to improve the quality of care and patient experience while reducing the costs. 5G also works for personalized and preventive care. (AT&T, 2020)

▶ Critical Infrastructure

The critical infrastructure refers to the systems people rely on, and if they were incapacitated or destructed, it would cause a debilitating impact on the physical or economic security, public health or safety. If any of the components forming the critical infrastructure fails, it may reduce the quality of people's life and even jeopardize it. Some examples of critical infrastructure are power grids, energy and water distribution, and telecom networks. It is thus of utmost importance that these components are highly protected against natural disasters, cyberattacks and other threats.

The characteristics of 5G make it suitable base for monitoring critical infrastructure. 5G can thus help public safety, governments, and utility companies to ensure the normal functioning of the services relying on critical infrastructure.

Ericsson has investigated the benefits of 5G on critical infrastructure and presents examples in their use case white paper. (Ericsson, 2020) As an example, energy and water utilities may connect to a large number of networked 5G devices while performing real-time, autonomous decisions.

The 3GPP has also investigated the use cases related to the critical infrastructure in the study on scenarios and requirements for 5G access technologies. (3GPP, 2019) This publication summarizes the special aspects for the public safety and emergency communications as well as public warning and emergency alert systems. Some examples of these are direct device-to-device (D2D) communications such as ProSe (3GPP, 2018) as well as Mission-Critical Push-to-Talk (MCPTT), Mission Critical Video transfer (MCVideo), and Mission Critical Data transfer (MCData) within those special areas. (3GPP, 2017)

The mission critical mode of 5G (and LTE) networks provides the public safety operators ways to use communications on-network using the 5G (or LTE) network infrastructure, or off-network, or in combined mode. This ensures the needed priority and available capacity for the very special environments.

In addition, 5G is planned to provide mechanisms to enable emergency calls including positioning and location for emergency calls, as well as Multimedia Priority Services. 5G also includes public warning services for notifications to users as per the regional regulatory requirements

▶Aviation and Drones

5G can work as a feasible platform for completely new verticals such as drones. They are developing vastly and can perform increasingly complex and autonomous tasks such as packet delivery. They benefit from large, fast and low-latency communications and control links.

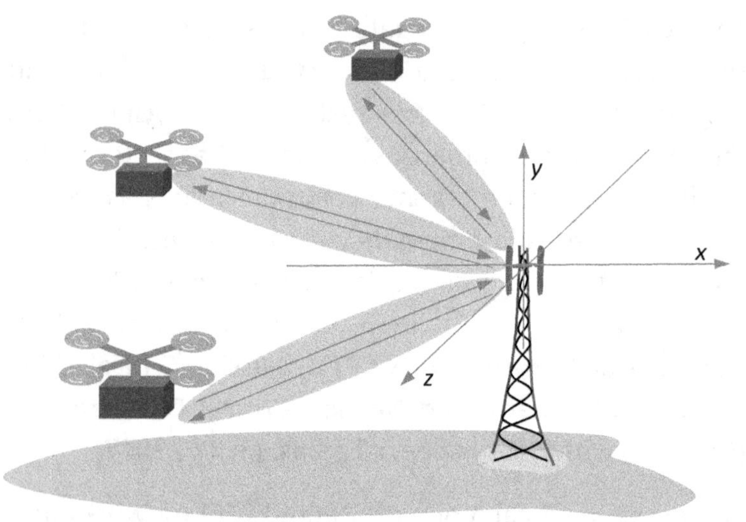

Figure 18 The 5G network can provide feasible control and communications channel for drones by offering enhanced, 3-dimensional adaptive coverage.

While the traditional cellular radio network planning tends to focus on ensuring the adequate coverage on the street level, the adaptive beam-forming antennas can serve well the new environment because the drones may require functional link also above the base station antennas. As an example, the GSMA and Global UTM Association (GUTMA) have explored ways in which the GUTMA is working to unlock the full value of connectivity in drones by the telecommunications industry. (GSMA, 2019)

Also, the commercial and special flights can benefit from the use cases the 5G networks provide. As an example, an airplane, as soon as it has landed, can upload the telematics data from the previous flight into the servers via eMBB network slice. The high speed at that very moment is essential because oftentimes, the commercial flights stay on ground only short time period while there may be vast amount of telematics data from the previous flight that needs to be backed up to the ground station. Meanwhile, the plane may need to receive files, software packages and other important contents via the URLLC network slice. The integrity of this data may be of utmost importance.

▶ Virtual Reality

Augmented Reality (AR) and Virtual Reality (VR) represent a special use case. In fact, the AR/VR use cases map oftentimes with a variety of other verticals.

AR means that a real-world view merges computer-aided view. There may be several types of human sensors involved such as visual, auditory and haptic.

VR, in turn, builds up the virtual environment. It can be imaginary space or based on real environment, and human beings can interact within this space. The underlying technologies include 360-degree panoramic video or light field forming advanced holographic views. There are already various AR/VR applications such as games, broadcasting, simulated education, virtual healthcare, and machinery assembly and maintenance training, to mention some.

The AR/VR data processing requires high-performance processors. The processing can be done by stand-alone hardware such as on a dedicated equipment or powerful desktop PC. Alternatively, the processing can also be offloaded onto a cloud environment. 5G could potentially provide a highly functional component for the latter by offering a communications link which transfers the raw AR/VR data onto cloud for postprocessing, and delivers it back to the 5G device. (Kelvin Qin, 2018)

▶Other verticals

The list above only scratches the surface of the possibilities for 5G to benefit verticals. There are much more verticals that may be able to function more efficiently using 5G services, such as retail and financial services (in-build security of 5G, platform for mobile ID), entertainment (low latency, fast data transfer for advanced multimedia), and agriculture (support for a massive amount of intelligent rural sensors working autonomously for users), to mention some.

As an example of the efforts to understand the needs of the verticals is the EU's 5G-PPP which is funding several projects, including vertical use-cases. (5GPPP, 2020) Another example is the GSMA which is coordinating research on the Network Slicing attributes best benefitting variety of verticals. (GSMA, 2019)

How the 5G Network Functions?

Even if we users do not really need to care about the call procedures beyond our devices, it is interesting to know what happens "behind the curtains".

The 5G system consists of two segments which are the radio and core networks. In between, there is a transport network that delivers the calls and messaging throughout the infrastructure.

The radio network connects smart phones and other 5G devices to the base stations within their radio coverage area. The 5G radio has been evolved from the LTE networks and supports much more capacity nowadays.

The core network connects the data calls with other mobile users within their respective radio coverage area—whether they are using 5G or earlier mobile generations—and services residing on data networks. One example of this is a server delivering video streaming.

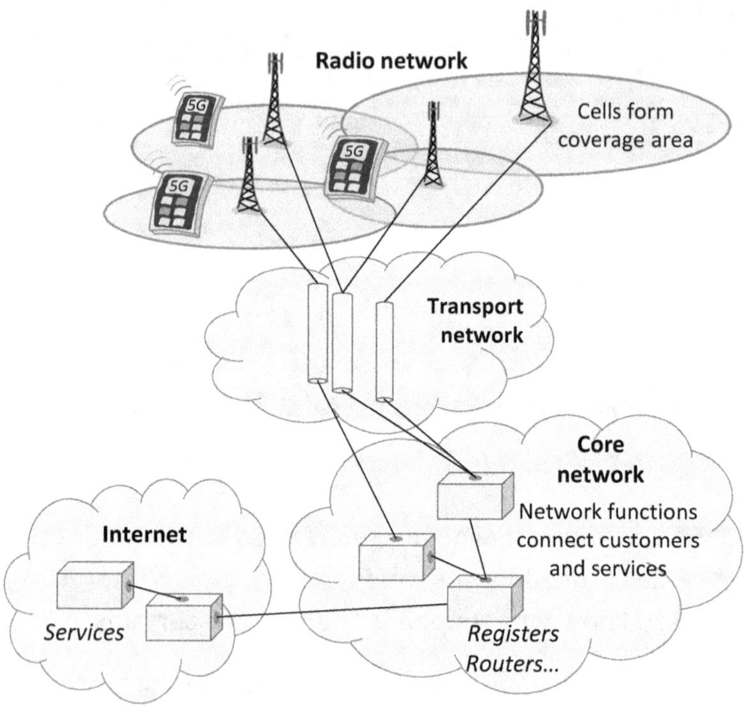

Figure 19 The 5G system consists of the radio, transport and core networks that make the connections possible between the 5G devices and services.

Radio Network

The radio network consists of base stations[33]. They are equipment shelters housing the transmitters, receivers, power supplies and other devices required for the radio connectivity with the mobile devices. The shelter can be a separate building, a room within already existing con- struction, or a simple box installed on a wall or pole.

---

[33] Technically speaking, we use a term "next generation Node B", or "gNB" when we refer to the 5G base station's radio cell.

The 5G base stations form individual radio coverage areas, cells. The cells are capable of interconnecting the 5G mobile devices and the 5G network so that the voice calls, messages, and data transfer, as well as the supporting signaling, may take place in both directions.

The cells are the foundation of the radio network so the mobile network operator needs to design their capacity and coverage adequately.

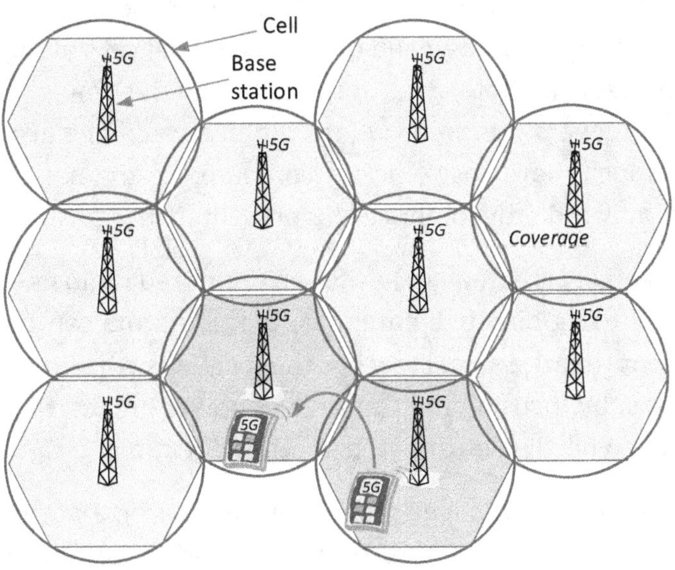

Figure 20 5G is a cellular network which means that the 5G base stations form coverage areas. In an ideal network, the cells overlap a bit so that the handover of the connection between cells is fluent for the moving subscribers. This ensures there is no service breakdown when we are on the move.

In an ideal radio network, there is always a bit overlapping between the cells so that the handover of the calls can take place without losing the connection when we move within the radio network coverage area. In practice, the large and smaller cells are layered further.

The 5G radio network may have a big number of cells. In a large area, such as USA or China, there may be tens of thousands of base stations for merely basic coverage, and each of them can form one or more cells. Furthermore, there may be a number of small, very low-power cells in a similar manner as the Wi-Fi hotspots to offer high-capacity locations. As an example, it makes sense to build up a tower by a highway and design individual cells along the route for the both sides of the tower. In this way, the operator covers the areas where the users actually are. This is much more cost-efficient compared to the use of a uniformly radiating omni-cell, as presented in Figure 21.

The directivity is actually a key for an optimized radio network. By using directive antennas, the operators can be confident that the signal reaches the locations where the users are, by focusing the transmitter power into certain direction while it is attenuated in other directions.

There are many directive antenna types ensuring that the operators may design their radio network accordingly case-by-case. In 5G, the radio network may support even further developed forms of the directive cells thanks to the adaptive antenna solutions. Such antennas are able to create the focused beams as per need for the active devices. This enhances the coverage and also makes the 5G base stations much more energy efficient.

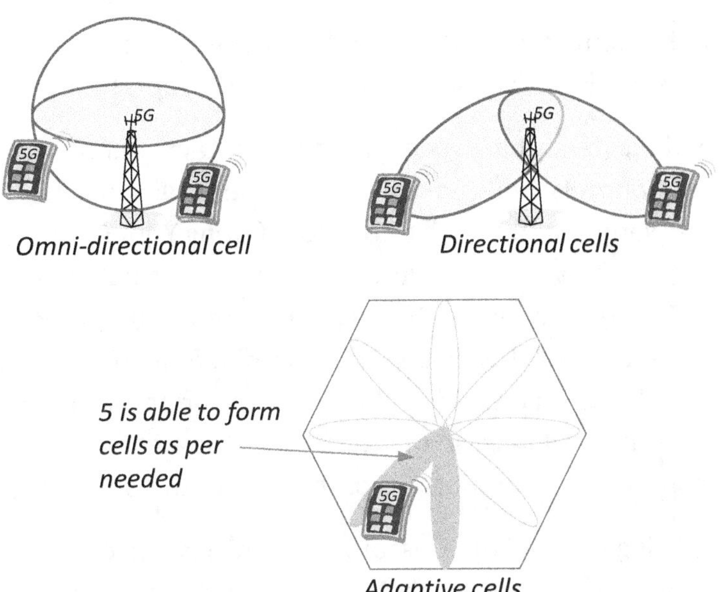

Figure 21 The 5G networks can support adaptive cells (beamforming) that optimizes the base station's power consumption.

In 5G context, the term "band" refers to a proportion of the frequency spectrum. Mobile network operators need to obtain at least one band in order to provide their cellular services to the consumers.

The band as such is quite transparent topic for us consumers, with one important exception; our 5G device needs to support at least one of our operator's bands in order to work appropriately.

There are no issues if you purchase your new 5G smart device at the retail store of the same operator that provides the contract for the 5G cellular services. Some operators have a set of certified devices based on their own requirements for the device vendors to guarantee the 5G device works as expected within the own operator.

The thing gets a bit trickier when you obtain your 5G mobile phone elsewhere and want to use it with a certain non-related operator. The concept of Bring Your Own Device (BYOD) is not new. In fact, the very original idea of the GSM back in early 90s was to ensure the transportability of the mobile devices and common functions of the networks. So, in the case of 5G, the same principle applies technically, but with so many options, it is a good idea to ensure the compatibility beforehand with the operator.

Assuming our favorite operator already has deployed 5G, and we are happy with their service level and radio coverage, we are ready to select the phone.

At this time, the support of the frequency bands plays a key role. 5G bands are a commonly and globally agreed set of possible blocks of the radio frequency spectrum which the operator can use in their business to offer the cellular service to us consumers.

The highest authority of all, ITU (International Telecommunications Union) dictates which frequency bands should be available in different parts of the world. ITU divides the world into three areas: Americas, Europe/Africa and Asia. The bands typically differ a bit between these areas, but there are also common bands in all regions. Please refer to Table 3 to get an idea of the most typical common bands worldwide if you want to ensure your 5G phone is truly global when roaming.

Core Network

The other, equally important part of the 5G is the core network. The core network houses all the needed functionalities for establishing, maintaining and releasing data and voice calls, and for establishing any other communication links such as messaging between the users and the network's own signaling.

The new 5G core network differs from the "traditional" architectural models significantly. Previously, all the network functions were handled by dedicated, operator-owned elements for those specific functions, such as subscription authentication or user register. These network devices were stand-alone, individual machines with their own hardware and software.

5G is based on virtualized network functions. Each and every function is thus merely a set of software instances that run on a common, virtualized hardware. This means that instead of having to use a mesh of the traditional and individual network elements, the 5G core networks can be operated in clouds of operator or 3rd party. Cloud typically refers to the resources offered by data centers.

These virtualized networks enhance greatly the performance of 5G compared to the previous generations.

Other novelty of 5G is the edge computing. It means that the operator can move the desired contents nearer to the user within the core network. If the user wants to see YouTube video, instead of the user contacting the video server somewhere in the other side of the network, the operator may want to transfer the contents to the network's edge nearby the user. It reduces the delays when the user keeps downloading and buffering the video on the go.

Another benefit of the edge is that the device could offload part of its processing onto it. That would be useful in, say, virtual reality contents creation and playback using a terminal that sends the raw data to the edge for its processing, and receives the "digested", post-processed data back to be merely displayed.

Those devices require much less processing power enhancing the battery life and lowering the cost of the device. If this concept gets popular, we could start seeing low-cost multimedia devices equipped with 5G connectivity. Example of such device is a simple visor to display 3D contents of games, movies and environments mixing augmented reality.

Some of the key tasks of core network are the establishment of the voice and video calls, and the delivery of messaging, data and related signaling. The 5G standards ensure that these functions are applicable both in the national as well as international environment.

The interoperability is thus an important aspect. The 5G networks need to comply with it so that customers can enjoy fluent user experiences in their own home operator's network as well as when roaming other countries.

The interoperability is ensured by the globally approved 3GPP standards.

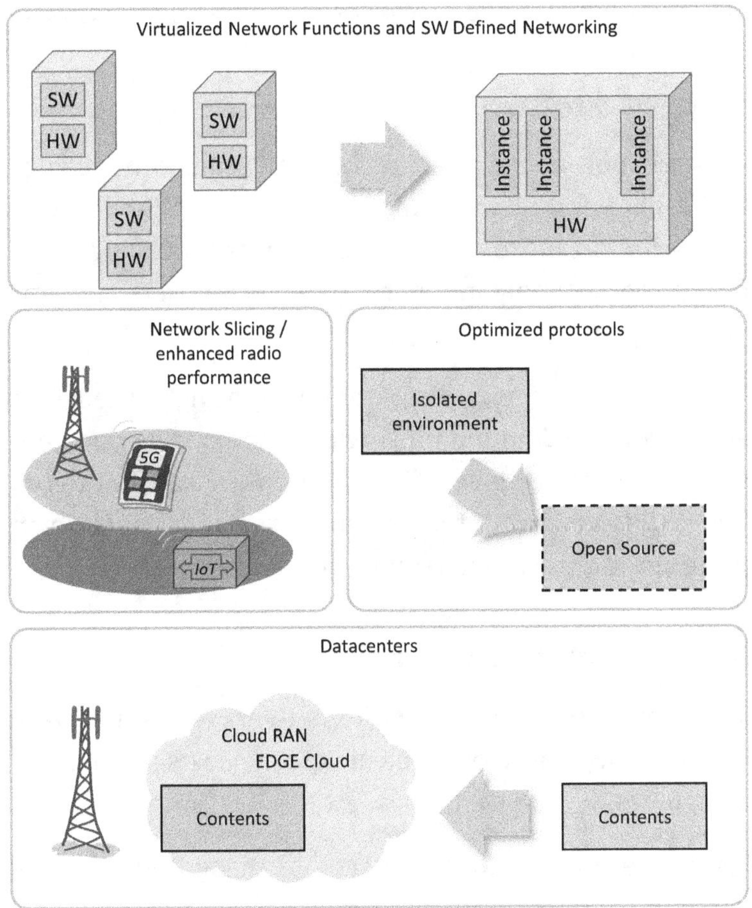

Figure 22 Some of the key differentiators of the 5G networks are the network virtualization (making the functions decoupled from the underlying hardware), network slicing (offering special networks over network), open source (opening doors for new ecosystem stakeholders) and edge (data centers).

The 3GPP ensures the 5G standardization and its evolution.

There are also supporting efforts for providing additional guidelines on top of the 5G standards. Some examples are the voice calls and messages as defined by the GSMA.

Transport Network

The transport network is a set of fiber optics, other cables and radio links between the radio and core networks. It has an important role to ensure the data is delivered correctly and in timely manner between the users and service providers.

The transport network is getting increasingly intelligent along with the demand for service based quality and geo-redundant service levels. The modern transport network benefits from the virtualization of the functions which can optimize the management of the transport.

Network Planning Principles

When operator plans the radio network, there is a very elemental key principle in balancing the costs as depicted in Figure 23.

If the operator compromises any of these items, coverage, capacity, or quality, the cost is lower but the customers will be unhappy for the service. In the worst case, the customer may decide to cancel the cellular plan to select and alternative operator.

This moving of the customers is called churn, and it is the worst case for any operator. It does not only decrease the profit—we customers are, after all, the principle generator of the money flow to the operators through our cellular plans—but the churn may also trigger critical messages on social media and potential degradation of the operator's brand loyalty.

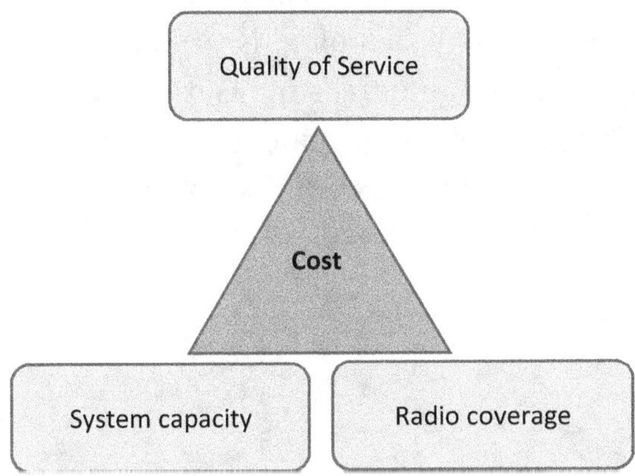

Figure 23 The cost of the cellular network depends on the designed capacity, coverage and quality of service.

It is said that the negative customer feedback reaches people 20 times more efficiently than the positive, so the operators really need to be aware of the development and ensure the churn is minimized by offering high-quality services and customer care.

On the other hand, if the operator over-dimensions the network by offering overwhelming capacity or coverage, which we consumers may not fully utilize within the near future, it is a waste of money and impacts negatively the business of the operator. In a good network, there is sufficiently overlapping to ensure us users with adequate quality, but not too much to keep the costs down.

So, the essential task of the mobile operators is to optimize the triangle of capacity, coverage and quality of their networks in order to maximize the return of investment yet making us consumers happy with the service.

This chapter only touches the high-level principles of the 5G network architecture and planning. If you are interested in learning more on the details of 5G, the **5G Simplified** (Amazon 2019) gives an introduction to the technical base including the overall idea of the new 5G architecture and functionality.

Is 5G Sufficiently Well Protected?

5G standardization teams have renewed the security model of the new networks. The security model incudes many novel solutions within the 5G network infrastructure such as multi-layer key generation based on the PKI (Public Key Infrastructure) concept. Figure 24 presents some of the aspects relevant to the renewed 5G security models.

✓ New operator's internal user identifiers are SUPI (Subscription Permanent Identifier) and SUPI (Subscription Concealed Identifier). The user keeps utilizing the established form of the telephony number.

✓ Protection of the contents and integrity of user data and signaling over the whole 5G network.

✓ Protection of the access to the functional core elements; only the elements that are allowed to request information from their counterparties can do so.

✓ Renewed security functions and network elements for forming and storing multiple security keys which are used for the subscribers and network elements.

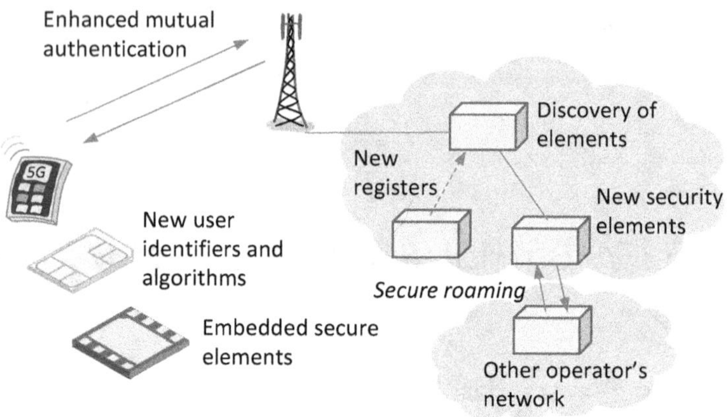

Figure 24 The security of the 5G networks is updated.

Also, it can be assumed that as 5G networks mature, the concept of embedded and integrated SIMs will be increasingly popular. They are beneficial for saving energy and space of the devices, and they can also be protected in a new way as the elements cannot be detached from the equipment.

The 5G security is transparent and automatic for us users. Nevertheless, it is always a good idea to follow generic protection guidelines such as keeping the mobile device's software up to date at all times, to avoid installing suspicious apps that request too much permissions, and to apply common sense in the utilization of apps in the endless task of avoiding phishing attacks and viruses from fraudulent links, etc.

How to Make Business Out of 5G?

5G can be assumed to generate plenty of new businesses. It is based on novelty solutions such as cloud-based networks and virtualized network functions. As a result, it also opens doors for many new stakeholders. One example of them is the new data center service providers.

Operators

The traditional business model for the mobile network operators (MNO) must consider the investment in the network infrastructure in order to offer adequate cellular connectivity for the customers. The most important Capital Expenditure (CAPEX) items include the radio and core network equipment and transmission in between.

The MNO needs a frequency license to operate the radio network. The license is typically valid for a finite time aligned with the expected life-time of the respective mobile communications network. The license needs to be purchased which happens via frequency auctions that the national government's frequency authorities coordinate. As the frequencies are limited, it has turned out to be rather lucrative business for the governments, and generates oftentimes billions of dollars per auctioned frequency band. (Fletcher, 2020) The auctions may sometimes raise the price of the most desired frequency block so high that some potential 5G operators might not have resources to invest in it.

The core network elements are expensive, too. A part of the radio and core network can be leased from the ones having the license. Such operators renting the radio capacity and redistributing it to their customers are referred to as Mobile Virtual Network Operators (MVNO). The liberation of the telecommunications back in 1990s has opened room for more competition, and the MVNO business model can be sustainable. As an example, there are dozens of MVNOs in the USA with a total of tens of millions of customers. This model can be expected to generate growing revenue as 5G takes off.

The virtualization of the core functions means that instead of building up the whole core network, the operator may also consider renting the functions and capacity from the various data center and cloud providers.

On the other hand, the subscription fees generate the principal profit of the operators. They thus need a sufficient customer base which is happy with the offered connectivity service's quality and cost.

In order to keep their customers satisfied, the MNO must build adequate radio coverage and offer sufficiently capacity. The optimal quality of service (QoS) is a key in order to minimize the churn. The customers abandoning the original operator and establishing a new contract with a competitor is one of the worst scenarios of the operators; the best way to avoid it is to make the services attractive both technically and cost-wise.

Although the network deployment costs vary greatly, a single base station (with a tower antenna for covering a sub-urban area) can easily cost 100,000 US dollars or more. To provide a nation-wide radio coverage, the cost of the radio network plays thus a major role and may require a multi-billion investment in dollars.

In addition to the investment in radio and core network elements, the MNO also has to maintain the network. The Operating Expenditure (OPEX) refers to the repeating costs related to the transmission infrastructure, power feeding, software and hardware updates, and replacement of faulty units, to mention only few. The OPEX items also include the support for SIM card supply as the cards need to be updated every now and then, pre-provisioned, stored physically and delivered to their destinations. Also, the management of possible subsidized mobile phones add up to the total operating costs.

In other words, the deployment and operation of a country-wide cellular network is a very expensive task.

The Return of Investment (RoI) of the MNO depends largely on the gain generated from the subscriptions. It is rather straightforward to estimate the revenue based on the customer base and average price of the cellular plan; for an operator with 10 million customers paying 50 dollars monthly subscription fee, the yearly revenue is 6 Billion dollars. The MNO can also consider supportive revenue-generating businesses which may or may not be related to the cellular services. One example of this is a bundled television broadcast service which could take advantage of 5G infrastructure, too.

5G also opens doors for new operational models. Examples of these are the Open RAN and Core. Instead of purchasing the network elements from traditional equipment manufacturer, the operator can obtain the needed functions from open networks in a vendor-agnostic way.

Another business model is related to the private network operations. Whereas the traditional cellular networks have required a license that the operators need to obtain prior to offering their connectivity for the customers, the 5G opens new opportunities for unlicensed band operators. These networks function on shared spectrum such as 3.5 GHz, and the specifications define concrete solutions for building up such system. In the USA, the CBRS (Citizens Broadband Radio Service) facilitates the sharing of the radio network among priority and best-effort users from which the latter ones do not need to invest to the license. These types of networks are highly useful in limited areas such as factories and remote areas.

Media

Media is one of the business areas which may benefit from the enhanced 5G capabilities. As an example, the edge computing of 5G offers, among other benefits, improvements in localization and personalization of the contents that current media platforms lack. The related business model for media could be to offer wireless, highly localized and personalized news channels for the 5G customers.

Combining added value services and the interactivity of the 5G, this model could interest customers to start utilizing services based on monthly plan or as-per-need. The innovation is thus in a key position to develop and deploy the attractive, useful and entertaining solutions.

One area of development may relate to the further innovation of the MBMS (Multimedia Broadcast Multicast Service). It has been defined originally already for 3G networks but the business models have not sufficed for its large-scale deployment. Thanks to the increased capacity of 5G, the MBMS mode may finally turn out to be successful commercial-wise.

The MBMS refers to the ability of the network to broadcast the same contents over defined areas for all the users desiring to subscribe to such communication channels. This shared approach saves greatly the capacity yet it provides means to personalize the user experiences via interaction channels which the user may utilize during the reception of the common contents.

Regardless of the technical mode the operator decides to use for the media, a realistic business model is required for maintaining a sustainable ecosystem. As the previous trials with the MBMS have demonstrated, this is not easy task. The more specific model may be still under development, but 5G may finally provide a highly feasible platform for media. The task is to find sufficiently attractive services from which the customers are willing to pay.

One of the still somewhat unexplored areas is the edge computing which may offer novelty ways for media to personalize the contents. Basing on the edge, the media could offer highly localized news and entertainment for different user segments within defined service areas.

Equipment Manufacturers

The device manufacturer, whether it is about network elements or mobile phones, needs to balance the costs and revenue. The cost items include the research and development, manufacturing, selling, and distribution of the equipment such as smart devices, mobile IoT equipment, vehicle-based devices, and many advanced components.

The optimization of the Bill of Material (BoM) plays a key role in the equipment business. One example of this balancing is related to the support of 5G frequencies in the smart devices. As can be seen from Figure 7 and Figure 8, the specifications define an overwhelming amount of bands to select from, but the more the device supports, the more complex, large, heavy, and expensive it gets due to the number of additional components such as antennas, filters and switches. For this reason, the mobile device manufacturers tend to select only the most useful bands and provides different models for different regions as can be seen in Table 6.

The optimization of the number of components is related to the energy consumption of the device, too. For us consumers, while the advanced functions are hidden within the mobile's electronics, their power consumption impacts directly on the user experience. The more advanced and numerous features, the less the battery tolerates it lowering the utilization time. With increasing pressures from the competitive field, the balancing of the features, performance and battery life is not at all easy task.

Especially for the network infrastructure and the 5G antennas, 3GPP (3GPP, 2019) places special emphasis on the balance of the complexity and performance. The analogue electronics has an important cost impact in the massive MIMO technology combined with advanced and active antenna systems. They are considered as an important component for optimizing the new radio of 5G. Thus, 3GPP recommends that the industry designs the respective requirements thoughtfully in order to enable economically viable deployments. Furthermore, 3GPP is required to support low-cost network infrastructures, devices, and operation and maintenance to enable economically viable deployments for the provision of minimal voice and data services for areas with very low level of Average Return Per User (ARPU).

Ecosystem

The "million-dollar question" for all the stakeholders, established and new ones alike, is how to figure out and develop realistic, sustainable business models for 5G.

The benefits of 5G include vastly increased bandwidth and intensive edge compute operations which equals to significantly higher data speeds, higher number of simultaneously communicating consumers and IoT devices, and faster response times. These are some of the key justifications for building up the 5G networks as they provide clearly better performance for the users compared to the previous mobile communications systems.

Nevertheless, it is important to understand that while the cellular networks keep enhancing the customer experiences by lowering latency and increasing the reliability, the long-term trend of the telecommunications industry indicates that the customers are assuming to "get more and pay less".

This commonly known phenomenon is seen especially in terms of reducing subscription fees. The development may, on fact, even limit the near-term feature development to merely faster downloads unless the industry figures out new applications and solutions that can better benefit from the 5G performance which, in turn, ensures future sustainable business.

All the stakeholders on 5G business have the same type of challenges and opportunities.

Developers

App development is one of the concrete means we consumers can get involved in the 5G ecosystem's business. 5G is still lacking the ultimate apps that can best take advantage of the high performance of the new system so the whole ecosystem can benefit greatly if more app developers start creating 5G-specific solutions.

The previous LTE networks can tackle many of the high bitrate -related solutions but the lower latency and higher data speed of 5G are still largely unexplored areas regardless of many ideas to develop more advanced apps such as Augmented Reality and Virtual Reality games. For this reason, there is plenty of room for the developers to contribute to the 5G ecosystem.

The operator community, in fact, may want to expedite the development of the new solutions that rely on the 5G networks. One of the related initiatives aiming to motivate the developer community is the 5G Operator Platform of the GSMA. Its goal is to create a common set of principles and tools such as APIs (Application Programming Interfaces) that can be utilized to expose the capabilities of the 5G infrastructure such as edge computing. By collaborating, the operators could offer a unified platform to federate multiple operators' edge computing infrastructure to give application providers access to a global edge cloud to run innovative, distributed and low latency services through a set of common APIs. (GSMA, 2020)

Consumers

The above-mentioned stakeholders maintain the techno-economic balance of the ecosystem. Thanks to the standardized 5G system and increasingly open environment, it is easier for new stakeholders to onboard and make business. An example of these new players are the cloud providers. This evolution, combined with open source concept and virtualized network functions, ensures that we consumers can start take advantage of the resulting scale of economies.

Like the previous mobile communication generations demonstrated, the consumers start favoring new capabilities of the advanced technologies when they provide sufficiently attractive benefits. In the initial phase of 5G, merely the data speed will increase thanks to the eMBB mode. That is good for such use cases as fast downloading of a 4k movie. That may not be sufficiently important reason for some of the current LTE users, to purchase a new smart device and to upgrade the subscription.

In a bit longer run, the individual consumers and businesses alike can benefit from the 5G performance, including evolved voice and data communications, and more efficient services such as fast, lag-free remote access.

The Open Source concept of 5G is one of the benefits for the users; the more stakeholders join the ecosystem, the greater variety of services there will be which keeps the cost balance at adequate level.

Is 5G Harmful for Us Users?

The social media and Internet house plenty of discussions on the potential impacts of 5G on our health, the opinions ranging from one extreme to another. So, what is the latest knowledge on the field and should we be worried about it?

Background

Observing the considerable growth of the mobile communication systems since the beginning of the wireless era back in 1980s, we are not able to see correlation with increasing malignant tumor cases on the population regardless of some sources suggesting the risk.[34] One might wonder thus where the doubt originates from.

5G is not the only mobile generation that has triggered such discussions; the same debate has taken place along

---

[34] *https://www.cancer.gov/about-cancer/causes-prevention/risk/radiation/cell-phones-fact-sheet*

with the deployment of the previous ones, too. It could in fact be that the word *radiation* causes the doubt. Uncontrolled radiation, as known from the nuclear power plant accidents, can result in severe health problems, fatal injuries or loss of lives when people are exposed to it long enough. Some examples of these problems include radiation burns, cancer and mutation of cells.

So, are there any similarities between the nuclear power or other dangerous sources of radiation and mobile communications systems such as 5G?

According to the scientific consensus, the short answer is no; they do not have anything to do with each other. For a bit longer answer, let's check some of the key aspects.

What the RF Radiation Is?

The cellular systems are based on the electromagnetic radio waves over the air as depicted in Figure 25.

All the wireless communication systems rely on the radio waves, including 5G, the difference being on how systems embed (modulate) the information on the carrier waves. Whereas the early Morse code was based on simply switching the analogue transmitter's carrier power on and off, the modern methods use much more advanced, digital methods.

The proportion of the frequency spectrum that is useful for all kind of wireless communications spans from few kilohertz (kHz) up to a few hundreds of Gigahertz (GHz).

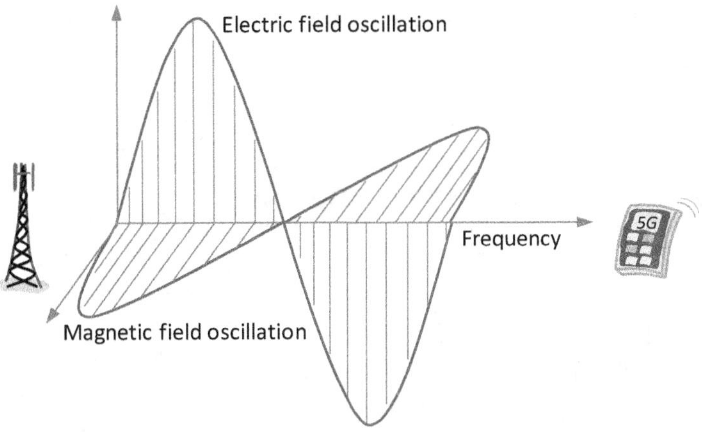

Figure 25 The principle of electromagnetic wave. The information is embedded into these carrier waves by modulating them.

This area is referred to as a radio spectrum while the frequency spectrum beyond it consists of visible light followed by ultraviolet, X, gamma, and cosmic rays as depicted in Figure 26.

All the wireless telecommunication systems are allocated within the radio frequency (RF) spectrum. The international and national frequency authorities plan the frequency allocation rules so that the wireless systems can work on their dedicated slots in harmony.

The ITU (International Telecommunications Union) dictates the global principles whereas the national frequency regulators make the country-specific plans. As an example, FCC (Federal Communication Commission) decides in the United States which commercial systems can use certain bands of the radio frequency spectrum. The detailed allocation table can be downloaded from the FCC.[35]

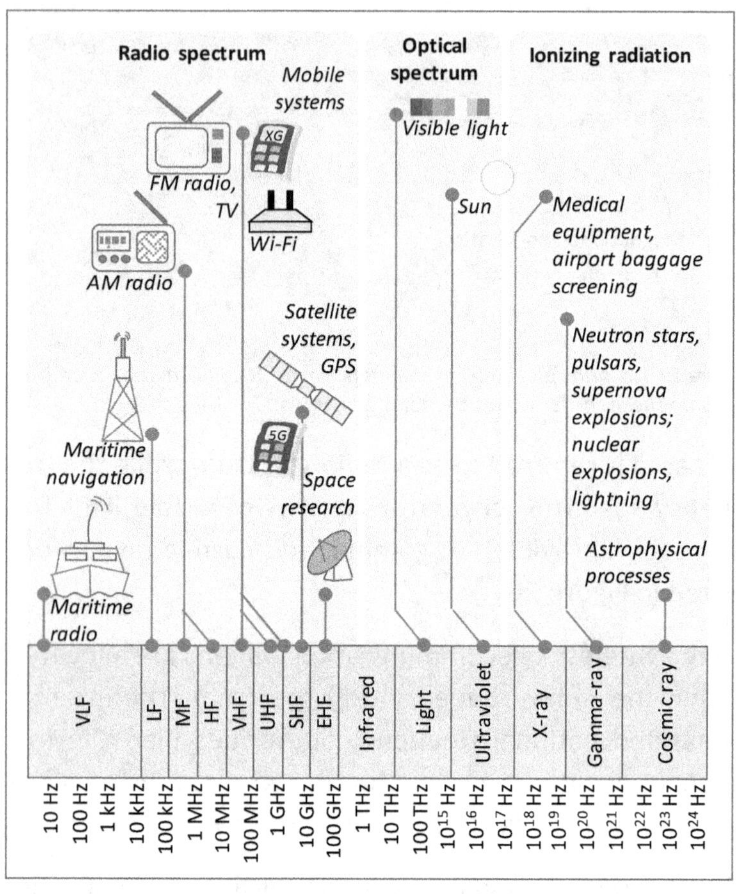

Figure 26 The spectrum and its typical uses.[36] Please note that the presented frequency scale is logarithmic. Thus, the frequencies of the range where ionizing starts taking place are about 100,000 times higher than on 5G bands.

---

[35] *https://transition.fcc.gov/oet/spectrum/table/fcctable.pdf*
[36] *https://www.nasa.gov/directorates/heo/scan/communications/outreach/funfacts/txt_band_designators.html*

As for the mobile communication systems, the regulators typically sell these proportions by coordinating frequency auctions. The national mobile network operators (MNO), such as Verizon, AT&T, T-Mobile, and many regional ones in the USA, bid licenses for certain period of time in order to offer their cellular services to us customers.

The Rules

In addition to the regulation of the system allocations, the frequency authorities also ensure that the transmitted power level and thus the RF radiation of the telecommunications equipment—such as smart phones and base stations—do not exceed the safety values. These values are set well below the commonly agreed limits for the exposure of the RF radiation that would start causing thermal effects on humans.

So far, the consensus of the scientific community is that the only proven effect of the non-ionizing radiation on us humans is the increased temperature of the biological tissue such as skin if the power level is high enough. So, in order to minimize this thermal effect, the limits for the radiating power of any radio transmitter are set so low that it is in practice impossible to even measure the difference.

Equally, there are strict rules for the authorized telecom maintenance personnel within the immediate proximity of the transmitter antennas e.g. on a tower. The minimum allowed distance depends on the radiating RF power level, and if there is a need to go closer, the systems within that range must be switched off meanwhile. On high towers, there may a variety of antennas of different systems such as cellular base station with a typical effectively radiating power of few tens up to few hundreds of watts in front of the antenna while some systems such as television broadcasting may have up to 50,000 watts of radiating power in front of the antenna.

Regardless of decades of studies and the broad understanding that the radio frequencies within the safety limits are not harmful for people, there are still questions about the potential health issues of the radio communications. This is understandable as we are not capable of observing the radio waves by human senses.

Challenges of the Research

So, as mobile phones are oftentimes used rather close to the head, could they then trigger any other health issues such as tumors?

It is not an easy question as the investigation of such event needs to consider any other potential sources such as previous or current exposure of a person or other subject on radon gas, X-rays, CAT (Computerized Axial Tomography) scan treatments, asbestos, environmental carcinogens, toxic chemicals, or naturally occurring genetic or other diseases such as a cancer that develop regardless of the external reasons and spread into the brain.

Of many studies, to understand their relevancy, we also need to understand any limitations and the impact of generalized assumptions, including the inaccuracies and errors due to statistically limited number of subjects or their differences with humans.

As an example, the U.S. Food and Drug Administration (FDA) nominated RF radiation for study by the National Toxicology Program (NTP) about the potential health effects of long-term cellphone use.[37] The activity studied rats and mice that were exposed to full-body RF radiation on frequencies that are used in 2G and 3G systems.

For the highest radiation, which exceeded four times the maximum regulatory limit when scaled to humans, the results indicate that there was "clear evidence" of tumors in the hearts of male rats, and "some evidence" on tumors in the brains and in the adrenal glands of male rats while it was unclear whether there was correlation between cancer and RF exposure for female rats, and male and female mice. For the levels that were comparable to the maximum allowed value for the humans, there were no health effects found.

The NTP scientists were not able to explain why male rats appear to be at greater risk for developing tumors compared to female rats—so the question remains whether it was due to RF exposure or other reason. In addition, NTP found longer lifespans among the exposed male rats

---

[37] *https://www.niehs.nih.gov/health/materials/cell_phone_radiofrequency_radiation_studies_508.pdf*

and reasons that this may be explained by an observed decrease in chronic kidney problems that are often the cause of death in older rats.

The source clarifies that the findings in animals cannot be directly applied to humans because the exposure and duration were greater than what people may receive from cellphones, and because the rats and mice received RF radiation across their whole bodies instead of localized exposures humans may receive. The study did not find correlation with the heath issues caused by the RF radiation on humans, but claims to have been able to question the assumption that radio frequency radiation is of no concern as long as the tissues are not heated.

With the open questions and inconsistencies that the abovementioned effort and some other studies leave, it is no wonder why the topic generates still confusions as indicated by the National Cancer Institute, USA.[38]

Levels of the Non-Ionizing Radiation

Before checking further statements of the scientific community, let's have a look at the principle of the radiation. Ionizing radiation, originated from such sources as X-ray and CAT scan equipment, has enough energy to detach electrons from atoms or molecules ionizing them. As a result, such ionizing radiation can harm human cells. If we

---

[38] *https://www.cancer.gov/about-cancer/causes-prevention/risk/radiation/cell-phones-fact-sheet*

were exposed too long time to the ionizing radiation, it can severely harm us, causing cancer and even death.

The mobile communication is always based on non-ionizing radiation which has not enough energy to impact atoms or molecules. Strong RF radiation originated from high-power radar etc. can cause thermal effects on living tissue, though, and that's why regulators have put such demanding limits for the wireless equipment.

The mobile industry follows the strict safety limits regulators have set for the RF radiation of mobile devices and base stations. As an example, although our home Wi-Fi router can work on the same band as microwave oven, it would not cause thermal effects as its transmitting power is so low. As a result, all the mobile device manufacturers need to comply with the maximum radiating power and SAR (Specific Absorption Rate) limits. The SAR is defined as the amount of energy absorbed by a mass of a biological tissue. The maximum allowed SAR value falls typically in the range of 1.6 – 2 W/kg. As an example, FCC has set the limit to 1.6 W/kg in the USA. Equally, the regulators put strict limits for the base station radiation.

As depicted in Figure 27, the power level of the 5G base station lowers fast as a function of the distance according to the free space loss formula.[39] Doubling the receiver's

---

[39] $P_r = P_t + G_t + G_r + 20\log\left(\lambda/(4\pi d)\right)$ where P_r and P_t are the received and transmitted powers in dBm, G_t and G_r transmitting and receiving antenna gains in dBi, λ the wavelength in m, and $\pi=3.14$.

distance from a transmitter means that the electro-magnetic field strength at the respective new location is reduced to one-quarter of its previous value.

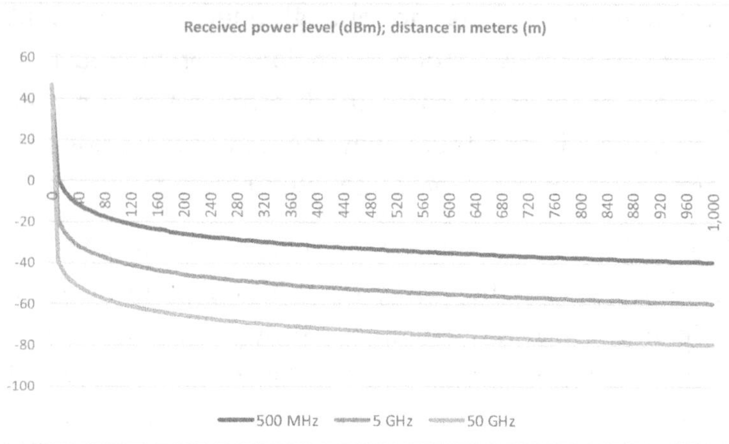

Figure 27 The received power from the base station lowers very fast the farther it propagates. On the other hand, the mobile device radiation is set to a very low level to comply with the SAR limits. This example presents the received power (in dBm) for a 50 Watt (47 dB) base station in freely propagation line-of-sight environment. In practice, there are obstacles such as buildings and vegetation that attenuate the signals further.

The maximum defined 5G base station radiating power is +47 dBm (50 W) for a medium area type, and +33 dBm (2 W) for a local area type.[40] This example demonstrates the fast attenuation of the low, mid and high bands under the

---

[40] As per the 3GPP Technical Specification TS 38.104 for the 5G base stations, the medium area type refers to Micro Cell scenarios with a minimum distance between the base station and the device along the ground of 5 m whereas the local area type represents Pico Cell scenarios with the minimum distance along the ground equal to 2 m.

most ideal conditions when no obstacles are found between the antennas, presenting the base station's maximum effectively radiating power of 50 watts (corresponding to +47 dBm, that is, decibels compared to one milliwatt) on 500 MHz, 5 GHz, and 50 GHz.

As an example, at 5 meters from the 50 W transmitter antenna, the received power is +6.6 dBm which equals to about 6 milliwatts (thousands of a watt) on the lowest presented, best propagating frequency of 500 MHz while it is a fraction of microwatt (millionth of a watt) on the highest presented frequency of 50 GHz. Depending on the band and frequency, the received power of about -80 to -95 dBm is still sufficient for lowest data connectivity.

It can be noted that that after the immediate proximity, the values are inferior compared to the maximum transmitted power of 5G handsets of +23 dBm (0.2 W).[41]

At the local areas of the city centers, a Pico Cell scenario can be used to provide the small cells. These base stations offer small hotspots of 33 dBm (2W), and their antennas can be thus installed e.g. on a light pole or on a wall. As a comparison, the wireless Wi-Fi routers support power levels of 20-23 dBm (100-200 mW) of indoor use cases up to 36 dBm (4 W) of the Broadband Wireless Access (BWA); the values depend on the country and band.[42]

---

[41] As per the 3GPP TS 38.101-2 for the 5G user equipment.
[42] *https://wlan1nde.wordpress.com/2014/11/26/wlan-maximum-transmission-power-etsi/*

Due to the low power and the radio propagation characteristics on 5G bands, the overall RF radiation of 5G base stations remains thus low. This is also the understanding of the industry, based on the available field tests such as the one Telstra Exchange has presented.[43]

Takeaways and Recommendations

Coming back to the question about the impacts of 5G on health, the available results, such as the previously mentioned project of the FDA, do not seem to provide a credible foundation for making science-based recommendations for limiting human exposures to low-intensity RF fields. As an example, the Canadian regulator states the following:

"Despite the advent of numerous additional research studies on RF fields and health, the only established adverse health effects associated with RF field exposures in the frequency range from 3 kHz to 300 GHz relate to the occurrence of tissue heating and nerve stimulation (NS) from short-term (acute) exposures. At present, there is no scientific basis for the occurrence of acute, chronic and/or cumulative adverse health risks from RF field exposure at levels below the limits outlined in Safety Code 6. The hypotheses of other proposed adverse health effects occurring at levels below the exposure limits outlined in Safety

---

[43] *https://exchange.telstra.com.au/5-surveys-of-5g-show-eme-levels-well-below-safety-limits/*

Code 6 suffer from a lack of evidence of causality, biological plausibility and reproducibility and do not provide a credible foundation for making science-based recommendations for limiting human exposures to low-intensity RF fields."

> *—The Minister of Public Works and Government Services, "Limits of Human Exposure to Radiofrequency Electromagnetic Fields in the Frequency Range from 3 kHz to 399 GHz; Safety Code 6".*

As the GSMA indicates further, expert groups and public health agencies such as the World Health Organization (WHO) broadly agree that no health risks have been established from exposure to the low-level radio signals used for mobile communications. Based on experience with 3G and 4G networks and the results from 5G trials, the overall levels in the community will remain well below the international safety guidelines.

Also, compliance assessment of 5G network antennas and devices are dictated by international standards which include new approaches for smart antennas and the use of new frequency ranges.

To be on the safe side, though, some operators state that individuals who are concerned about radiofrequency radiation exposure can limit their exposure, including using an earpiece and limiting cell phone use, particularly among children.

For the ones interested in learning more details on the topic, it is a good idea to seek for high-quality and non-biased studies carried out by expert entities that rely on the scientifically adequate methods and principles.

Table 12 summarizes some sources of information to ease the research.

Table 12 Resources for additional information on health aspects of the mobile communications.

Source	Description
Joint Venture Silicon Valley	***Bridging the gap***; 21st Century Wireless Telecommunications Handbook. (Joint Venture Silion Valley, 2020)
World Health Organization	***Establishing a dialogue on risks from electromagnetic fields***. (WHO, 2019)
US National Cancer Institute	***Cellular phones and cancer risk*** (National Cancer Institute, 2019)
Health Canada	***Safety Code 6, 2015*** (Health Canada, 2018)
Swisscom	***5G Mobile Technology Fact Check*** (Swisscom, 2019)
GSMA	***Safety of 5G Mobile Networks*** (GSMA, 2019)
ITU workshop on 5G, EMF & Health	***RF exposure impact on 5G rollout***: A technical overview, Warsaw, Poland, 5 December 2017 (ITU, 2017)
IEEE C95.1-2019	***IEEE Standard for Safety Levels*** with Respect to Human Exposure to Electric, Magnetic, and Electromagnetic Fields, 0 Hz to 300 GHz (IEEE, 2019)

Frequently Asked Questions

In addition to the previously presented topics, we customers may have still many doubts on 5G. The following presents some of the questions 5G has generated.

Please feel free to comment and discuss further the 5G development on the author's LinkedIn page and blog:

▶ *www.linkedin.com/in/jypen*

▶ *www.5g-simplified.com*

How 5G differs from 4G?

5G is a renewed mobile communications system that provides highly evolved radio and core networks. Comparing the 5G and previous generations, the main differences of the radio are the support of higher frequency bands and wider bandwidths. 5G has technical means to increase the data rates and lower the latency, too. This makes the 5G radio network feel more fluent to the users.

The core network has experimented a total face lift, too. The new architecture is based on virtualized network functions which share a common hardware for the tasks 5G needs. In the older model, the operators have separate, individual equipment and their own software for each functionality within the network. The virtualization makes 5G more efficient.

As a consequence, the evolved 5G network can provide up to 10-20 Gb/s data speeds and down to 1 ms latency values, combined with ultra-high reliability serving the most demanding applications.

Is 5g really needed?

The communication needs of the consumers, businesses, verticals and other users have increased steadily. The LTE networks can deliver an impressive capacity and their service areas have developed considerably so it is justified to question the need for yet another generation.

Nevertheless, as has been the case with all the previous generations, the increased performance of the new networks enables novelty applications and services that set gradually the new level of expectations by their users. 5G is capable of offering more fluent user experiences for many use cases such as remote education and virtual training.

Not only the technological performance is important per se, but 5G is expected to contribute to the overall development of the society cross nations. 5G is expected to generate new jobs and provide a variety of direct and indirect business opportunities to many new stakeholders within the telecommunication industry and beyond it.

Yet another aspect is the role of 5G in increasing people's security and wellbeing. The new networks are formed by many types of cells of which the dense small cell areas are in a special position. As one of their benefits, they can provide enhanced location accuracy information in special cases such as medical emergency, expediting the first responders to find the patience thanks to the increased accuracy of the horizontal and vertical cellular-aided positioning solutions enabled by the customer.

Continuing the benefits, 5G small cells can help in the lifeline communications by offering network infrastructure's redundancy if, say, natural disaster or other issue damage it.

Do the higher 5G frequencies harm us?

5G will be relying increasingly on higher frequencies on millimeter waves that densifies the network. One might thus wonder if the higher frequencies or the denser 5G network harm us.

As the GSMA Mobily Policy Handbook[44] states, the World Health Organization (WHO) and the International Telecommunication Union (ITU) recommend that governments adopt the radio-frequency exposure limits developed by the International Commission on Non-Ionizing Radiation Protection (ICNIRP), reviewed and updated in 2018.

Observing the overall picture from Figure 26, and the more specific table of the FCC solely for the US frequency band allocations[45], we can see that the radio spectrum has been rather occupied already for long time with numerous systems relying on the electro-magnetic radiation and producing RF emissions. The dense 5G small-cell deployments on the millimeter waves are located within the same spectrum between the current cellular frequencies and already deployed satellite and other systems. They all are well below the spectrum of the visible light, which in turn is well below the highest frequencies where the ionizing range starts to be a concern.

As described in Chapter "Is 5G Harmful for Us Users?" the current consensus of the scientific community is that there is no correlation between the RF radiation of cellular networks or devices and health issues. With 5G being added into the safe range of the spectrum, in the non-ionizing area, the industry is ensuring the total RF power levels would not exceed the regulated limits.

---

[44] *https://www.gsma.com/publicpolicy/mobilepolicyhandbook/*
[45] *https://transition.fcc.gov/oet/spectrum/table/fcctable.pdf*

For the ones interested in understanding better the overall landscape, the following resources provide a common-sense explanation:

- ✓ *The New York Times: **The 5G Health Hazard That Isn't**.*[46] This source investigates one of the root sources and reasons for the controversy statements.

- ✓ *CNBC: **Is 5G Safe?*** CNBC has made a common-sense, impartial report on the safety of 5G.[47] This YouTube video presents the scientific consensus on safe 5G.

Will Wi-Fi 6 make 5G obsolete?

Both Wi-Fi 6 (IEEE 802.11ax) and 5G share a same foundation. Both provide enhanced data speeds and low latency performance.

They do not compete against each other, though, but they can work in a parallel fashion as complementary technologies. 5G provides mobility within its deployed coverage area while Wi-Fi 6 continues providing the familiar, fixed hot spots.

The latter keeps serving as a wireless router at home and office in the same way as the current Wi-Fi devices do. In fact, the 5G specifications support the fluent and seamless interworking between both technologies.

---

[46] *https://www.nytimes.com/2019/07/16/science/5g-cellphones-wireless-cancer.html*
[47] *https://www.youtube.com/watch?v=Ag1hkv2Upww*

More information on the topic can be found from the summary of Cisco.[48]

Can 5G protect us against cyber world?

The cybersecurity protects us users and communications infrastructure against cyberattacks. 5G can play an important role in this ecosystem.

Many functions within the telecom networks, including self-optimizing network concept, rely on artificial intelligence (AI), and it will be embedded increasingly into a variety of devices. It provides great benefits to make life easier but it also has raised some serious concerns about the safety issues. As an example, Prof. Steven Hawking indicated his doubts over the potential failure in controlling the machines that may eventually outsmart us. (Post, 2014)

"Success in creating AI would be the biggest event in human history. Unfortunately, it might also be the last."

—Steven Hawking, Huffington Post, 5 May 2014

Despite some of the action movies that present future machines outsmarting and attacking humans, we are still on rather safe side of such development. Nevertheless, many of us would agree about the importance to agree

---

[48] *https://www.cisco.com/c/m/en_us/solutions/enterprise-networks/802-11ax-solution/nb-06-5-things-WiFi6-5G-infograph-cte-en.html*

international rules for the further development of the AI to avoid machines taking over the control in the decision making related to our wellbeing (or to threatening it).

On the other hand, the AI has huge potential to make our life easier and safer, and to help businesses perform more efficiently in their daily tasks, so it is highly unlike that the further advances of the AI would suddenly stop.

Even if 5G has been designed primarily to provide us consumers with evolved connectivity and a platform for evolved services and applications, the role of 5G could be even more versatile. One future idea for the developers of the evolving system could be to provide means on how 5G could monitor and control the IoT environment, and steer the AI-equipped machines to perform their tasks adequately also in case of unexpected failures that, in the worst case, could otherwise lead into potential hazards for the people.

This topic goes clearly into ethical and philosophical side, but could trigger discussions in the international forums dealing with the future technologies and well-being of the people. There are already entities considering the ethical AI problematics and principles. An example of such environment is the AI-equipped, self-driving vehicle and the problematics on how it can minimize the damage in case of inevitable accident—which are not easy items to deal with.

For the interested ones, the study paper "Ethical Dilemma of Artificial Intelligence and its Research Progress" by Lei Ma et al[49] presents some of the ethical issues brought about by the application of technology and references for the further studies.

---

[49] *https://iopscience.iop.org/article/10.1088/1757-899X/392/6/062188/pdf*

5G Terminology

This section presents some of the most common terms related to 5G we consumers may face in our efforts to understand how to use the system.

3GPP (3rd Generation Partnership Project) is a standards development organization (SDO) defining the technical specifications of 5G. (***www.3gpp.org***)

5G refers to the fifth generation of mobile communication systems. The formal criteria for mobile communication systems, in order to be called "5G", are set by ITU (International Telecommunications Union) and documented in IMT-2020 (International Mobile Telecommunications) requirement set.

Band is a generic term for the proportion of radio frequency. Each 5G operator has one or more frequency bands which are used for radio transmission and reception. Operators typically purchase the right to use frequency bands for their operations from auctions organized by national regulators. In 5G, the low-band refers to spectrum below 1 GHz, mid-band to 1-6 GHz, and high-band to spectrum range above 6 GHz.

Base Station refers to a physical site where radio and transmission equipment are located for interconnecting user equipment (UE) and cellular network. In 5G, the radio component is gNB (next generation Node B). In LTE, it is eNB (evolved Node B), and in 3G it is NB (Node B).

CA (Carrier Aggregation) refers to the combining of radio capacity of two or more radio frequency bands for a single user. CA has been applied since 2G, but 5G is able to benefit more from it, thanks to the wider bandwidths.

Core Network connects the 5G base stations and delivers the data traffic and signaling between the user equipment and destination such as other mobile user equipment or server. In 5G, the core network is based on virtualized architecture, clouds, and data centers.

EMF (Electro Magnetic Field) is a common term referring to the radiation properties of radio transmitters, and it is related to safety regulation and health aspects. While EMF refers to radio waves influencing living tissues, the **EMC** (Electromagnetic Compatibility) is applicable term for indicating the influence of radio waves to equipment.

eSIM is a "virtual" version of the traditional SIM card. It means that the functionality of the SIM card is embedded into the user equipment such as smart device. As some of the special devices such as sports trackers are small, the removable SIM might not fit into it. Thus, there are new SIM variants which can be soldered and integrated into the hardware of the device. Furthermore, there are new solutions to manage the subscription over the air instead of switching the SIM between devices. RSP (Remote SIM Provisioning) of GSMA is one of these methods.

Licensed Band refers to the radio frequencies operators buy to get right to use them for their customers. National regulators typically auction licenses to be used for limited time period.

MIMO refers to Multiple In, Multiple Out antenna type which is used to transmit and receive via multiple radio paths enhancing the data speed. The more the inputs and outputs (radio paths) the antenna has, the higher data speeds can be achieved.

Millimeter Wave, or mm-Wave, is the term for the new bands on higher frequencies than were deployed in previous systems. 3GPP defines Frequency Range 2 for mm-Wave bands whereas Frequency Range 1 is for sub-6 GHz.

NFC, Near Field Communications, is a radio communications technology for a very short distance, 4-10 cm between the devices. It is meant for only small-scale information exchange so it can be used for contactless payments, as well as a means to set up other type of communications channel such as Bluetooth audio by tapping the device close to the activation point.

NR refers to New Radio which is the new 3GPP standard for the 5G radio system. It enhances the radio performance providing faster data speeds, more capacity and lower delay (latency).

NSA, Non-Standalone 5G network, is based on intermediate 5G architecture model which reuses part of the legacy 4G infrastructure. By using NSA in the initial phase, operator can expedite the offering of basic 5G services while constructing the fully capable, more performant 5G core.

Radiating Power is the basic measure of the 5G base stations' and mobile devices' transmitted power. For the 5G base stations, the radiating power level is in range of 43-47 dBm (decibels compared to 1 milli Watt, typically used in telecom industry) which equals to 20-50 Watts.

As a rule of thumb, the highest-power 5G base stations have tower-installed antennas which are typically directive, meaning that they have a certain, e.g., +10...+17 dB gain compared to omni-radiating antenna. Once the signal is transmitted and propagates from the antenna, the received power level decreases exponentially as a function of the distance between the transmitting and receiving equipment. To ease the respective power calculations, the decibels are used in telecommunications.

As an example, if the radiating power at the base station's transmitting antenna is 50 dBm (100 Watts), the frequency is 1 GHz, and the distance between the base station and cellular phone is 100 meters, the path loss may be around 90 dB depending on the conditions. There typically occur additional path losses due to obstacles such as trees and buildings. In this example, the typical received power level at the mobile phone's antenna may be about +50 dBm - 90dB = -40 dBm, which equals to 0.1 microwatts (fraction of one millionth of watts) applying typical values and urban path loss model. This value, at the same time, represents still sufficiently good signal strength for a successful data reception. If you are interested in testing more different values, there are tools available in Internet, e.g., at the following address: ***https://5g-tools.com/5g-nr-link-budget-calculator/***).

The presented example above applies to a large, macro cell which provides rather good coverage on rural areas. The radiating power of the 5G small cells, in turn, are comparable to the Wi-Fi home routers, meaning that their radiating power is significantly lower. The respective antennas can be installed on light poles and walls of the buildings.

SA, Standalone 5G network, is the full architectural version of 5G, including its own new radio and core networks. SA is the ultimate goal of operators as it is able to offer the maximum performance without the limitations of legacy 4G.

Spectrum Sharing refers to techniques to optimize the radio network capacity by letting more than one provider to offer the same band for their customers. 4G has already solutions such as SAS (Spectrum Access System) and US-based CBRS (Citizens Broadband Radio Service). There is also further optimized method introduced in 5G called Dynamic Spectrum Sharing (DSS), which changes automatically the capacity from the same band among 4G and 5G users, and LSA (Licensed Shared Access).

Unlicensed Band does not require licensing fees, and they can be shared among many providers. The downside is that the provider of such band cannot guarantee the service level because of varying capacity demands of different users. Examples of unlicensed band solutions include CBRS (Citizens Broadband Radio Service), LAA (Licensed Assisted Access), and NR-U (New Radio of 5G in Unlicensed mode as of Release 16).

Use Case refers to a specific situation in which a product or service could potentially be used. Moreover, a use case is a methodology in system analysis to identify, clarify, and organize system requirements.

Abbreviations

3D   3-dimensional

3GPP   3rd Generation Partnership Project

5G   Fifth Generation mobile communication system

5GC   5G Core network

5GS   5G System

A2DP   Advanced Audio Distribution Profile (Bluetooth)

AI   Artificial Intelligence

aptX   A codec capable of wirelessly transmitting 24-bit high-resolution audio

AR   Augmented Reality

BDS   Chinese positioning system

BT	Bluetooth
BYOD	Bring Your Own Device
CA	Carrier Aggregation
CAT	Computerized Axial Tomography
CDMA	Code Division Multiple Access
C-IoT	Cellular IoT
CN	Core Network
CPE	Customer Premises Equipment
C-V2X	Cellular Vehicle to Everything
DLNA	Digital Living Network Alliance
DN	Data Network
DSDA	Dual SIM Dual Active
DSDS	Dual SIM Dual Standby
eMBB	evolved Mobile Broadband
EMC	Electromagnetic Compatibility
EMF	Electro Magnetic Field
eNB	evolved NodeB (LTE "base station")
eSIM	Electronic SIM ("virtual" SIM)
FCC	Federal Communications Commission
FM	Frequency Modulation (radio)

FR   Frequency Range

FWA   Fixed Wireless Access

Galileo   European positioning system

GLONASS Russian positioning system

gNB   Next generation Node B (5G "base station")

GNSS   Global Navigation Satellite System

GPS   Global Positioning System

GSM   Global System for Mobile Communications (2G)

GSMA   GSM Association

GSM-R   GSM for Railways

HD   High Definition

HDMI   High Definition Multimedia Interface

HO   Handover

HSPA   High Speed Packet Access

ICT   Information and Communication Technology

IMT-2020 International Mobile Telecommunications 2020

IoT   Internet of Things

ITU   International Telecommunications Union

LE   Low Energy (Bluetooth)

LTE   Long Term Evolution (4G)

MIMO   Multiple In, Multiple Out (antenna)

ML   Machine Learning

mMTC   Massive Machine Type Communications

MNO   Mobile Network Operator

MTC   Machine Type Communications

MVNO   Mobile Virtual Network Operator

NFC   Near Field Communications

NFV   Network Functions Virtualization

NGC   Next Generation Core (5G)

NMT-900   Nordic Mobile Phone

NR   New Radio (5G)

NSA   Non-Standalone

OS   Operating System

OTA   Overt the Air

OTG   On-the-Go (USB)

OTT   Over the Top

PKI   Public Key Infrastructure

QoS   Quality of Service

QR	Quick Response (code)
QSZZ	Japanese positioning system
RAM	Random Access Memory
RCS	Rich Communications Services
RF	Radio Frequency
RoI	Return of Investment
ROM	Read-Only Memory
SA	Standalone
SAR	Specific Absorption Rate
SIM	Subscriber Identity Module
SMS	Short Message Service
SoC	System on Chip
SSD	Solid State Drive (memory card)
SUCI	Subscription Concealed Identifier
SUPI	Subscription Permanent Identifier
SW	Software
UMTS	Universal Mobile Telecommunications System (3G)
URLLC	Ultra-Reliable Low Latency Communications
USB	Universal Serial Bus

V2V   Vehicle to Vehicle

VoLTE   Voice over LTE (4G)

VoNR   Voice over New Radio (5G)

VR   Virtual Reality

WHO   World Health Organization

Wi-Fi   WLAN, based on the Institute of Electrical and Electronics Engineers' (IEEE) 802.11 standards

WiMAX2   Worldwide Interoperability for Microwave Access (4G)

WLAN   Wireless Local Area Network (Wi-Fi)

WRC   World Radiocommunication Conference

Bibliography

3GPP. (2017, June 20). *3GPP*. Retrieved from Mission
    Critical Services in 3GPP:
    https://www.3gpp.org/news-events/1875-
    mc_services

3GPP. (2018). *TS 23.303: Proximity-based services.* 3GPP.

3GPP. (2019). *TR 38.913: Study on scenarios and
    requirements for the next generation access
    technologies (Release 15).* 3GPP.

5GAA. (2020). *V2X.* Retrieved from https://5gaa.org/5g-
    technology/c-v2x/

5GPPP. (2020). *5G and Verticals.* Retrieved from
    https://5g-ppp.eu/verticals/

AT&T. (2020). *5 ways 5G will transform healthcare.*
    Retrieved from

https://www.business.att.com/learn/updates/ho
w-5g-will-transform-the-healthcare-
industry.html

Comms Update. (2019, April 4). *Verizon launches mobile
5G in Chicago, Minneapolis; claims 'world's first',
ahead of South Korea*. Retrieved from
https://www.commsupdate.com/articles/2019/0
4/04/verizon-launches-mobile-5g-in-chicago-
minneapolis-claims-worlds-first-ahead-of-south-
korea/

Ericsson. (2018, December 18). *5G meets Time Sensitive
Networking*. Retrieved from
https://www.ericsson.com/en/blog/2018/12/5G-
meets-Time-Sensitive-Networking

Ericsson. (2019, May). *5G consumer potential: Busting
the myths around the value of 5G for consumers* .
Retrieved from
https://www.ericsson.com/498f26/assets/local/r
eports-papers/consumerlab/reports/2019/5g-
consumer-potential-report.pdf

Ericsson. (2020). *Critical services and infrastructure
control*. Retrieved from
https://www.ericsson.com/en/5g/use-
cases/critical-services-and-infrastructure-control

FCC. (2020, February 10). *The FCC's 5G FAST Plan*.
Retrieved from FCC: https://www.fcc.gov/5G

Fletcher, B. (2020, January 14). *FierceWireless*. Retrieved from U.S. millimeter wave auction reaches more than $7B: https://www.fiercewireless.com/5g/u-s-millimeter-wave-auction-hits-above-7-billion

Frost, C. (2019, August 16). *Business Insider*. Retrieved from 5G is being used to perform remote surgery from thousands of miles away, and it could transform the healthcare industry: https://www.businessinsider.com/5g-surgery-could-transform-healthcare-industry-2019-8

GSMA. (2018, June 14). *Requirements for multi SIM devices*. Retrieved from TS.37: https://www.gsma.com/newsroom/wp-content/uploads//TS.37-v4.0.pdf

GSMA. (2019, September 6). *5G for verticals: reshaping Europe's mobile sector narrative*. Retrieved from https://www.gsmaintelligence.com/research/2019/09/5g-for-verticals-reshaping-europes-mobile-sector-narrative/799/

GSMA. (2019, December 24). *5G Innovation*. Retrieved from Global 5G launches: https://www.gsma.com/futurenetworks/technology/understanding-5g/5g-innovation/

GSMA. (2019, July). *5G Spectrum*. Retrieved from GSMA Public Policy Position: https://www.gsma.com/spectrum/wp-

content/uploads/2019/09/5G-Spectrum-
Positions.pdf

GSMA. (2019, December 24). *Global forecast for RCS
growth*. Retrieved from
https://www.gsma.com/futurenetworks/rcs/glob
al-launches/

GSMA. (2019, December 23). *GUTMA and the GSMA
Announce Collaboration to Help Define the
Future of Aerial Connectivity*. Retrieved from
https://www.gsma.com/iot/news/gutma-and-
the-gsma-announce-collaboration-to-help-
define-the-future-of-aerial-connectivity/

GSMA. (2019, July). *Safety of 5G Mobile Networks*.
Retrieved from
https://www.gsma.com/publicpolicy/wp-
content/uploads/2019/06/GSMA_Safety-of-5G-
Mobile-Networks_July-2019.pdf

GSMA. (2019, 12 18). *The Mobile Economy 2018*.
Retrieved from The Mobile Economy 2018
Report:
https://www.gsma.com/mobileeconomy/wp-
content/uploads/2018/05/The-Mobile-Economy-
2018.pdf

GSMA. (2019, December 24). *The SIM for the next
Generation of Connected Consumer Devices*.
Retrieved from https://www.gsma.com/esim/

GSMA. (2020, January). *Operator Platform Concept.* Retrieved from Phase 1: Edge Cloud Computing: https://www.gsma.com/futurenetworks/resources/operator-platform-concept-whitepaper/

GSMArena. (2019, December 24). *Phone finder.* Retrieved from GSMArena: https://www.gsmarena.com/

Health Canada. (2018, 18 April). *Limits of Human Exposure to Radiofrequency Electromagnetic Energy in the Frequency Range from 3 kHz to 300 GHz.* (Health Canada) Retrieved July 4, 2019, from Safety Code 6 (2015): https://www.canada.ca/en/health-canada/services/environmental-workplace-health/consultations/limits-human-exposure-radiofrequency-electromagnetic-energy-frequency-range-3-300.html

Holland, P. (2019, June 25). *CNet.* Retrieved from AT&T 5G: Our tests yield the wildest speeds yet: https://www.cnet.com/news/at-t-5g-our-tests-yield-the-wildest-speeds-yet/

IEEE. (2019, October 4). *IEEE C.95.1-2019.* Retrieved from IEEE Standard for Safety Levels with Respect to Human Exposure to Electric, Magnetic, and Electromagnetic Fields, 0 Hz to 300 GHz: https://ieeexplore.ieee.org/document/8859679

ITU. (2015, September). *Recommendation ITU-R M.2083-0*. Retrieved from IMT Vision: https://www.itu.int/dms_pubrec/itu-r/rec/m/R-REC-M.2083-0-201509-I!!PDF-E.pdf

ITU. (2017, December 5). *RF exposure impact on 5G rollout A technical overview*. Retrieved from Presentation of Kamil BECHTA, Nokia Mobile Networks, 5G RAN : https://www.itu.int/en/ITU-T/Workshops-and-Seminars/20171205/Documents/S3_Kamil_Bechta.pdf

Joint Venture Silion Valley. (2020). *Bridging the Gap*. Retrieved from https://jointventure.org/images/stories/pdf/JVSV_Wireless-Telecommunications-Handbook_2ndEd_DEC2019.pdf

Kelvin Qin, M. Z. (2018, April). *GSMA*. Retrieved from Network slicing use case requirements: https://www.gsma.com/futurenetworks/wp-content/uploads/2018/07/Network-Slicing-Use-Case-Requirements-fixed.pdf

Mu-Hyun, C. (2019, December 3). *South Korea secures 4 million 5G subscribers*. Retrieved from ZDNet: https://www.zdnet.com/article/south-korea-secures-4-million-5g-subscribers/

National Cancer Institute. (2019, January 9). *Cellular phones and cancer risk*. (National Cancer

Institute) Retrieved July 4, 2019, from Cancer causes and prevention: https://www.cancer.gov/about-cancer/causes-prevention/risk/radiation/cell-phones-fact-sheet

Penttinen, J. (2015). EMF - Radiation Safety and Health Aspects. In J. Penttinen, *The Telecommunications Handbook* (p. 956). Wiley.

Quain, J. R. (2018, August 10). *Digital Trends*. Retrieved from Georgia is paving the way for a high-tech, sustainable highway: https://www.digitaltrends.com/cool-tech/the-highway-of-the-future-is-being-paved-in-georgia/

Swisscom. (2019, March 27). *5G Mobile Technology Fact Check*. (Swisscom) Retrieved July 4, 2019, from Press Release: https://www.swisscom.ch/en/about/news/2019/03/27-5g-mobile-technology-fact-check.html

WHO. (2019, July 31). *Electromagnetic fields (EMF)* . Retrieved from EMF: www.who.int/peh-emf/en/

Index